If Truth Be Known

This book is an examination of claims of scientific support for the views 1) that Earth is a recent creation, not more than several thousand years old, and 2) that much of the current topography of the Earth and many (or most) of the fossil-bearing rock layers in Earth's crust are the result of a cataclysm that includes a worldwide flood in recent time. The discussion within this book is permeated by the conviction that modern science is compatible with Christian faith.

Clarence Menninga

COVER PHOTO

Palouse Falls, Washington, USA

Photo Courtesy of Pacific Northwest National Laboratory

The height of the falls is 198 feet. The channel of the Palouse River is carved through several successive lava flows of the Columbia Basin Basalts. (See Chapter 15 for more details.) The contacts between successive lava flows are clearly seen; the lip of the falls is near the top of one flow, with an older, thicker flow below that. There is a more recent flow above and immediately to the left and right of the lip of the falls, and a thicker, still more recent flow above that one is visible beyond the stream channel, with some weathered remnants of that layer visible above and to the left of the lip of the falls.

ISBN: 098588231X
ISBN-13: 97809858823-1-0

DEDICATED TO

All my teachers

ACKNOWLEDGMENTS

I owe a huge debt of gratitude to my colleagues at Calvin College, especially to those Faculty members in the sciences, and those in theology and religion, both in the College and in Calvin Theological Seminary, for refining my understanding through many discussions of the relationships of science and Christian faith. I especially appreciate Howard Van Till, Davis A. Young, John Stek, Ralph Stearley, and Gerry Van Kooten for their influence in developing my ideas. I thank Ralph Stearley and Gerry Van Kooten in Geology and Susan Felch in English for reading the manuscript and making many valuable suggestions. I am grateful to those many people outside the College with whom I have held discussions and carried on correspondence on topics of science and Christian faith, including some who strongly disagree with my convictions regarding the history of the universe.

I thank Patricia Menninga for providing the hourglass pictured in Figure 6. I am grateful to my non-scientist wife Irene for her encouragement, and for reading an early version of the manuscript and identifying some areas that needed clarification and simplification.

CONTENTS

FORWARD

When I was a youngster growing up in rural Iowa in the 1930's and 40's, I heard about the bones of some strange creatures called dinosaurs being found in western North America, and I was curious to know more about them. But I was told by adults in my Christian community that "Those creatures never *really* existed; those stories are just the wild imaginings of godless scientists who are trying to lead Christians astray." (I have talked with others of my generation who grew up in Christian communities, and who were told that same thing.) When I grew older I found out that the bones are very real, and that those dinosaur creatures really *did* exist. Those scientists were telling the truth, and my fellow Christians were mistaken; that realization was a sharp disappointment for me.

I can readily find excuses for those fellow Christians of my childhood. I don't think they intended to mislead or misinform me. Perhaps they didn't really know whether dinosaurs actually lived on Earth or not; most of them had very little formal education, and they did not have ready access to the massive amounts of information that are made available to us in this 21st Century through television, the internet, and a host of books at all reading levels. Very likely they were simply repeating something they had read or heard somewhere, without checking to find out whether it was true or not.

There are many Christians in our contemporary society who are convinced that Earth is a recent creation, not more than several thousand years old. Some of our fellow Christians subscribe to a catastrophist interpretation of history, believing that our present world has gone through major, worldwide catastrophic events in recent time that have shaped what the world is like. Those beliefs are based on theology, that is, on what those Christians understand the

Bible, the Judeo-Christian Scriptures, to be teaching. There is no dishonor in those beliefs.

Some of the Christians who believe that Earth is a recent creation, and that worldwide catastrophic events in recent time are responsible for forming many aspects of our present world, have attempted to support those theological beliefs with arguments from the scientific study of God's world. Frequently, passages from the professional scientific literature, journals, and books are quoted in claims of such scientific support. In some cases, however, the data reported in the claim of scientific support for the theological belief are not the same as the data reported in the original scientific publication, that is, the data from the scientific source are not reported faithfully. In some cases only a part of the scientific story is presented, with important details having been left out. Some of the quotations from the professional scientific literature that have been used in supposed support of those theological understandings have misrepresented the conclusions of the scientific publication from which the quotation was taken. In some cases, respected scientific principles have been misunderstood, and the misunderstanding has led to mistaken conclusions. In some cases the most reasonable and most straightforward conclusions from the scientific observations have been rejected, and replaced with unfounded conjecture. This book is about some of those faulty claims of scientific support for the theological conclusions of Earth as a recent creation, and of Earth as the product of major, worldwide catastrophic events in recent time.

Many of these faulty claims for support of theological beliefs by the results of scientific investigation are passed along among Christians without careful examination to determine whether or not they are true. Should we also try to find excuses for fellow Christians who remain less than well informed about science in this 21st Century, as we might have for the Christians that surrounded me in my youth, who were mistaken about the existence of dinosaurs on Earth? Not all of us are trained in science to any appreciable extent, nor should we expect everyone to be a professional scientist. During the past few decades I have often been invited to lead some discussion on the relationship between science and Christian faith for adult study groups in Christian churches, and for groups of teachers in Christian schools. Among those groups I have found many fellow Christians who have some interest in the results of scientific investigation and implications of those results for Christian theology. They have some questions about conflicting stories they have heard or read somewhere, but they haven't had the training, either formally or by self-study, that would enable them to make informed judgments about which version of the story is true. One audience member commented to me publicly, "You say this, and so-and-so says that; you have a Ph.D., and he has a Ph.D., who am I supposed to believe?" Those people, those who attend Christian churches from Sunday to Sunday, those who have legitimate questions about conflicting stories from science and Christian faith, but have had little or no training in science; those are the people I have in mind in writing this book. If you are a member of that group, layman or clergy, I think that this book will be helpful to you.

We recognize that there are legitimate differences among Christians in our understanding of the meaning of various passages of Scripture that refer to our material world, as, for example, in our interpretation of the Creation narrative in Genesis chapters one and two. It is painful for me, however, to hear fellow Christians trying to support a particular theological belief with arguments and claims about our material world which I know to be incorrect or misleading.

Can we find a place for an honest and straightforward examination of such claims and arguments to find out whether or not they are true? Earl Douglass, the discoverer of dinosaur skeletons at what is now Dinosaur National Monument, was a teacher and a student of nature during the late 19th and early 20th Centuries. Growing up in a Christian family, he learned geology and something of the history of living organisms on Earth during his young adult years. In his diary entry of June 1, 1884 he wrote, "Had a dispute with Father and Nettie [his sister] tonight about evolution and the Bible. Such disputes are of no good use."

The family circle is not a good forum for carrying on such discussions. The church pulpit is probably not a proper avenue for examining whether such arguments and claims are true, either. Sad to say, disputes about such matters have sometimes alienated family members or church members from each other. In my judgment, the issues involved in such topics are not important enough to be pursued to the disruption of brotherly love in families or in churches. Nevertheless, we should find a place for such discussion; we are obligated as stewards of God's creation to search out the truth regarding such matters to the extent that we are able to do so.

This book is an attempt to bring out some of the many claims and arguments about our material world that have been circulated among Christians, and to examine those claims and arguments to find out whether or not they are true. I will try to use ordinary and common language as much as possible, but it will be necessary in that examination to use some technical scientific terms; I will try to define or describe those technical terms as simply as I can whenever they are used.

The discussions presented here are about ideas and claims. The claims and ideas that will be considered are taken from publications that promote the view that Earth is a recent creation, not more than several thousand years old, and from publications that promote the view that our present world has been shaped by recent, worldwide catastrophic events. The sources of those claims and arguments will be provided, as well as the sources of information that form the basis for evaluating the validity of the claims and arguments being considered. References to those sources are listed at the end of each chapter.

Some of the publications chosen as sources for ideas and claims considered in this book were published quite a long time ago. Nearly all of the ideas and claims considered in this book, however, continue to be published and distributed among Christians to the present time. Older sources that are used have been chosen because they state the claim more clearly, or more fully, or because claims from the older publications have been more widely distributed in the Christian community. In particular, the older book *Scientific Creationism* receives a considerable amount of attention because it has been advertised as "the most practical, well-organized resource handbook on scientific creationism today" on the back cover of the 1974 issue, and on the back cover of the 12th Printing in 1985. The book continues to be advertised as "authoritative and thoroughly documented" on the back cover of the 24th Printing, 2006, and continues to be offered for sale both as a print edition and as an electronic book edition.

The claims and arguments to be considered in this book deal with topics from the physical sciences, primarily geology, physics, and chemistry. The matter of biological evolution by natural processes is not addressed in this book; the defense of those ideas, or opposition to them, is left for other writers.

The specific topics discussed in this book cover quite a broad range of results of the scientific study of God's world. Several chapters in the latter part of the book will be devoted to a discussion of radioactivity and the use of radioactive isotopes for measuring the ages of things; it is important to devote some space to those topics, because a large number of published claims of scientific support for the view that Earth is a recent creation have focused on those topics.

We will, however, start with a chapter that deals with the nature of scientific investigation, we will conclude with a positive statement of faith in the Christian gospel message, and we will insert a positive look at an interesting part of God's world near the middle of the book.

The readers of this defense of truth might possibly conclude that the book is a personal attack on certain persons. But that is as far from the intent as east is from west. I do not make any judgments about anyone's motivations or character.

1 WHAT SCIENCE IS LIKE

The claims and ideas that we will consider in this book have to do with the results of the investigation of God's world by scientific methods. So we should begin with a brief discussion of what science is like.

How do we learn about God's world through scientific investigation? What kinds of things or ideas can science help us to understand? What kinds of things or ideas is science unable to help us learn? How confident can we be about scientific claims? How are scientific explanations tested and evaluated? How can the layman, someone not trained in science, get beyond "I believe it because my teacher told me so."?

Scientific investigation and scientific study are based on two very important convictions about the nature of God's world, namely:

1. The physical universe is an ordered reality, that is, it possesses and displays order.

2. We humans are capable of perceiving and understanding the order in the universe, at least in part.

All scientists, both Christian and non-Christian, subscribe to those two convictions. If we did not, there would be no point in pursuing scientific study.

Some people would call those two statements "suppositions" or "basic assumptions" because they cannot be established by logical proofs. But much of our experience persuades us that those two statements are true, so I prefer to call them convictions. We should note that those two convictions about the physical world fit comfortably within the Judeo-Christian concept of God as Creator, and humans as God's image bearers. We also note that many ancient and primitive societies did not think of the universe as ordered; their superstitions arose from a conviction that the universe is capricious and unpredictable.

The quest for scientific explanations begins with our observations. We notice some regularity or pattern, perhaps in the motions of objects in the sky, for example, and we seek to understand how such motions fit with the structure and behavior of what we see. Once an explanation has been suggested, we test and evaluate that explanation by making further observations; if the observations fit with the explanation, our confidence in that explanation increases. If the observations conflict with the suggested explanation, that explanation becomes less acceptable to us. In all cases, the confidence that we have in a scientific explanation depends on how nearly our observations fit with that explanation.

Observations of our world make use of our physical senses such as sight and smell, of course. In addition, we make use of instruments and devices that aid our senses, such as microscopes and telescopes. We also make use of instruments such as voltmeters and thermometers and magnetometers that can measure or detect various aspects of our world that are not directly accessible to our senses.

Because science depends on observations of physical entities such as matter and energy, scientific explanations are limited to matters pertaining to the physical world. Science helps us to understand the structure and behavior of the physical world, but it is unable to answer questions about purpose or meaning. We get answers to questions about purpose and meaning from our philosophy and our religion.

The results of the scientific study of the past few centuries have become embedded in our everyday lives. From early childhood we learn about such scientific near certainties as the spherical, rotating Earth, and the dependable march of the seasons as the Earth revolves around the sun. The "Tiger" comic strip in Figure 1 illustrates that scientific discoveries of the past have become common knowledge in our modern culture.

Figure 1. Common knowledge. TIGER © 1984 KING FEATURES SYNDICATE Reprinted with permission.

Like Tiger in the comic strip, we "could have told you" that the Earth goes around the sun, but we tend to forget that such ideas were based on countless observations and hard work, aided by sound logic, and that general acceptance of those conclusions as the preferred explanation was initially resisted by many of the respected authorities of the time. The history of science and scientific ideas is filled with examples of how we came to "discover" or develop explanations that are not at all obvious at first glance.

We should note, also, that it requires some diligent effort to pursue scientific understanding of our world, and it takes some hard study to learn about the results and the observational evidence on which those conclusions are based. Tiger, again, in Figure 2, on "simple" explanations:

Figure 2. Simple explanations. TIGER © 1985 KING FEATURES SYNDICATE Reprinted with permission.

Like Tiger, most of us tend to want simple and certain answers to our questions, but science doesn't readily yield to those desires. Hard work has enabled us to learn a great deal about our physical world through scientific study and investigation, but it isn't always easy or simple.

Albert Einstein commented (in German), *"Raffiniert is der Herrgott, aber boshaft ist er nicht."* Literally translated, "God is subtle, but he is not malicious." Or, paraphrased by Einstein himself, "Nature hides her secrets by her essential grandeur, not by her cunning." Uncovering those secrets requires hard work and sound logic, but the work will be rewarded with reliable understanding because God is not trying to mislead us by the structure he has created into the world.

We must remember that scientific investigation is carried on by humans, and scientific explanations are therefore always imperfect, and always incomplete. (I hasten to add, especially for the benefit of the theologians and philosophers who might read this, that the same is true for theology and philosophy.) If our scientific explanation is inconsistent with our observations, the explanation needs to be modified, or, in extreme cases, rejected and replaced with a better explanation. Scientific explanations are always open to review and evaluation, and always subject to change if the observations dictate the necessity of such change. Some people complain that "those scientists are always changing their minds" about this or that. But such change is an integral part of scientific understanding. If scientists didn't change their minds when they are faced with evidence that conflicts with the existing scientific explanation, we would still be thinking about the Earth as flat and stationary in space.

So, if science is human explanation based on human observations, we might ask, "Where is God in the structure and behavior of the physical world?" God is everywhere and in everything. John Calvin wrote, "We know God, who is himself invisible, only through his works."[1]

11

DENNIS THE MENACE

"I'M PLAYIN' CATCH WITH GOD. "...AND HE THROWS
SEE? I THROW THE BALL UP..." IT BACK!"

Figure 3. Playing catch with God. DENNIS THE MENACE
© 1985 HANK KETCHUM ENTERPRISES. NORTH AMERICA
SYNDICATE. Reprinted with permission

In Figure 3, a Dennis the Menace cartoon, Dennis understood God's presence in the world in exactly the right way.

We might ask, further, "Where is the recognition of God in the scientific explanation?" The explanations that we read and hear about in science do not explicitly mention God. Why is God's name and action not included, since there have been many Christians who were influential in developing those explanations?

The reason God is not explicit in the scientific explanations is that those explanations are based on our human observations of the physical world, and God is not explicitly visible in the structure and behavior of that physical world. There is not an explicit "factor" for God in the equations that describe the effects of gravity on a baseball, nor in any other scientific explanation.

Our faith in God does not depend on our scientific understanding of the physical world. Our faith is not jeopardized by well-supported results of scientific study, even when such results are different from what previous generations might have thought the world is like. "If you believe in your heart that Jesus is Lord, and confess with your mouth that God raised him from the dead, you will be saved." (Romans 10:9)

However, we must be careful to distinguish between ideas that arise from scientific investigation and ideas that are derived from religion and philosophy. We must be alert and discerning with regard to what we read in newspapers, magazines, books, and the internet, and with regard to what we hear and see on radio and television, because ideas from religion and philosophy are often mingled with ideas that arise from scientific investigation. Statements about the existence or non-existence of God, for example, do not come from science. The statement "The cosmos is all that there is, or ever was, or ever will be."[2] is a statement of faith, and not a conclusion based on observation.

We must also be on our guard so that we are not led astray by misinformation about God's world. There is misinformation being published in various places. Some of it is done in fun; in Figure 4, a Sunday comic strip of Calvin and Hobbes, Calvin and his dad are sitting on the steps watching a brilliant red sunset.

Figure 4. CALVIN AND HOBBES © 1989 Watterson. Dist. By UNIVERSAL UCLICK
Reprinted with permisssion. All rights reserved.

It's all in fun, of course. Calvin represents the curious and gullible small child, and his dad is the source of mistaken ideas (maybe knowing better, but having a little fun with Calvin), and then copping out when the inconsistencies in his perspective get exposed by questions about important details. I don't think any of the readers of this book would take the misinformation in this Calvin and Hobbes comic strip as serious science.

In our contemporary world, however, there is a good deal of misinformation being published as if it were scientifically reliable. The remainder of this book is about some of that misinformation.

We should add a note about what are sometimes called "values" in science, meaning widely accepted standards of performance of scientific investigation, and widely accepted standards of honesty in reporting and discussing the results of scientific investigation. These standards may be expressed in various ways, some more detailed than others, and I will provide only a partial list of a few that I think are of greatest importance.

1. Honesty: The scientific investigator must be scrupulously honest in reporting his/her observational data and measurements,

and in making inferences or drawing conclusions based on the observations and measurements. She/he may not "fudge" the data, reporting numbers or observations which are, in fact, not the numbers or observations that were gained in the investigation.

2. Completeness: The scientific investigator must report all of the observational data and measurements that were obtained in the particular study being considered. She/he may not omit some of the data just because it is not in agreement with the conclusions that he/she might have anticipated or desired. If a series of measurements is involved, and one or more of the measurements does not fit well with the inferences or conclusions drawn, valid reasons for omitting such a measurement from consideration must be provided.

3. Openness: The scientific investigator is expected to openly share observations and measurements with the wider community, especially the community of other scientists. Observations of the physical world are not generally considered to be private property.

NOTE: In today's world, there are two areas in which this openness is curtailed: 1) Proprietary information, sometimes called "company secrets," that provide some real or perceived advantage to the company in a competitive economy, to the extent of sometimes obtaining a patent on information; and 2) Government secrets that are guarded to protect the security of the nation. There is ongoing discussion about the proper extent to which openness may be curtailed in either or both of those cases.[3]

4. Faithfulness: Use of published or privately obtained data from other investigators must be faithfully reproduced and faithfully represented by the scientific investigator and/or reporter who is making use of those data.

It is common for scientific investigators to use data and/or conclusions drawn from other investigators, either to buttress one's own conclusions, or to argue for one's own conclusions as superior to those of the other investigator. However, 1) she/he may not modify a direct quotation from another source in such a way as to alter its meaning; and, 2) he/she may not lift a quotation or information from another source out of its context in the original source, and use it to convey a meaning other than that of its meaning in its original context. In summary, the investigator or reporter must honor the values of honesty and completeness in references to the work of others, as well as in his/her own work.

In later chapters we shall have occasion to call attention to cases in which these "values" in science have been violated.

References

[1] John Calvin, *Commentaries on the First Book of Moses called Genesis* (1554), translation by Rev. John King (Grand Rapids: Eerdmans, 1948), 59.

[2] Carl Sagan, *Cosmos* (New York: Random House, 1980), 4.

[3] Dorothy Nelkin, *Science as Intellectual Property: Who Controls Scientific Research?* AAAS Series on Issues in Science and Technology, (New York: Macmillan, 1984).

2 Frozen Mammoths of Siberia

Part or all of the carcasses of several woolly mammoths, extinct creatures very similar to elephants, have been found frozen in the permafrost of northern Siberia. One of those, known popularly as the Beresovka mammoth, has been claimed as evidence that Earth is a recent creation, and that recent worldwide catastrophic events have produced many of the features presently observed on Earth. That claim is mistaken, however, and the true story provides a different conclusion. We will focus attention on just two features of the story, 1) the identification of the undigested plant food that was found in the mouth and stomach of the animal, and 2) the opinions expressed regarding how rapidly the carcass had become frozen.

So here is the story. In 1900 a Siberian hunter came across a portion of a woolly mammoth protruding above the surface on the slope alongside the valley of the Beresovka River in northern Siberia, just south of the Arctic Circle. The carcass, embedded in permafrost, apparently had been uncovered by a soil slump on that slope. The find came to the attention of the municipal authorities of the region, and word was sent to the Russian Imperial Academy of Sciences at St. Petersburg. An expedition was immediately organized under the leadership of Dr. Otto F. Hertz [a.k.a. Herz] to recover as much of the frozen animal as possible. He was assisted by Eugen W. Pfizenmayer, the preparator and taxidermist of the Museum of the Imperial Academy. The body of the mammoth was excavated in 1901, cut into pieces for shipping to St. Petersburg, and reassembled in the Museum of the Academy.

The find caused great excitement, and a great deal of material has been published about that mammoth in the scientific and the popular literature. The initial report of the expedition to recover the mammoth was published in Russian in 1902.[1] Excerpts from that report were translated and published in English in a publication of the U.S. National Museum, Smithsonian Institution under the title *Frozen Mammoth in Siberia* in 1904.[2] Excerpts from the initial report by Herz were also translated and published in a book entitled *The Mammoth and mammoth hunting in north-east Siberia* in 1926.[3] Pfizenmayer published his own account of the expedition in the book *Siberian Man and Mammoth*,[4] an English translation of his book that had been published in German in 1926. We will make use of those first-hand accounts in evaluating claims of recent catastrophic cause of death, and sudden, rapid freezing of the animal.

The full report of the expedition was published as a 3-volume monograph (in Russian), with translated title *Results of the Imperial Academy scientific expedition to retrieve the Beresovka mammoth in 1901*; Volumes I and II were published in 1904, and the results of the scientific examination of the remains were published as volume III in 1914. Much of that monograph is available only in Russian, but excerpts from some parts have been published in English, and will be referred to later in this chapter.

The "stuffed" animal is on display today in the Museum of the Imperial Academy, in the pose in which it had been found, with about one-third of the skin and hair being the original tissue and the remainder reconstructed in proper taxidermic style. The skeleton of the mammoth is on display nearby, with one tusk added to replace a missing one, with broken bones repaired, and with a few missing parts filled in.

Many stories have been repeated about the animal and its excavation, some true and some fanciful fiction.

Mouth and stomach contents

The claim that the undigested plant material in the mouth and stomach of the mammoth was "tropical" or "sub-tropical" has been circulated widely. That story has been incorporated into some publications that promote the view that Earth is a recent creation to support the claim of sudden climate change in recent time in a catastrophist view of history. For example, in *The Creation-Evolution Controversy*:

> "The flesh of the Beresovka mammoth was edible, the stomach containing about 30 pounds of food consisting of subtropical vegetation still in an undigested state, and the mouth was filled with partially masticated food."[5]

A footnote to that discussion of frozen mammoths refers to "The Pre-flood Greenhouse Effect" in *Symposium on Creation II* in which several features of the Beresovka mammoth are listed, including:

> "The digestive tract: Within the digestive tract were found tender sedges, partly undigested, and partly still green. Twenty-seven pounds of vegetation, all of subtropical or mild temperate types (according to the best translations from the Russian) were found in one portion of the intestinal region."[6]

Such claims of "tropical" or "sub-tropical" food in the mammoth's stomach have been circulated among some of my personal acquaintances, also.

But that claim is not true. There was partly chewed plant food in the mouth of the animal, and incompletely digested plant food in its stomach. But the plants found there are not tropical, and not subtropical, nor even mild temperate.

Some of the plants that were found in the mouth and stomach of the mammoth were identified, and a list of the Linnaean classification of nine species of plants was published (in Russian).[7] The list consists of species of grass and sedge that are found growing in northern Siberia today, in the same region where the frozen carcass of the mammoth was found. That evidence confirms that the vegetation and the climate of the region during the mammoth's lifetime, and at the time of its death, was pretty nearly the same as it is today.

The conclusion that the plants found in the mammoth's mouth

and stomach are northern Siberian plants, and not tropical or sub-tropical, is supported by the following quotations from primary sources, that is, those that were written by the leaders of the expedition to recover the carcass of the mammoth, and from publications that followed quite soon thereafter.

The list of species of plants found in the mouth and stomach of the mammoth was translated and published by I.P. Tolmachoff, who was working in the U.S.A., but had access to the primary reports through personal friends in Russia. Along with the list, we find the following general comment:

"All of these species are typical representatives of a meadow flora of Northern Siberia at the present day. Leaves and branches of bushes were not found, although they had been not lacking on the shores of Beresovca. In summer the mammoth was a grass-eater who, like the recent reindeer, preferred this food to any other."[8]

The identical list was published in Henry F. Osborn's monograph on the Proboscidea. The following general comment accompanies the list:

"On examination this material proved to contain a flora of no great variety, but of exceeding interest, because it consisted of plants that are still native to the place, i.e., Beresovka River, northern Siberia. .... All these kinds of plants are found in the same locality at the present day and form a characteristic meadow flora."[9]

A more recent list of species of plants found in the mouth and stomach of the Beresovka mammoth, compiled and published by Farrand, consists of the nine species already mentioned, plus numerous species identified from pollen found with the plant material. The pollen analysis was attributed to a paper by Kupriyanova,[10] published in 1954. The pollen would have come from plants that were native to the region, but the plants themselves would not necessarily have been present in the mouth and stomach of the mammoth. The general comment accompanying the list is as follows:

"Stomach contents reveal an abundance of grasses, sedges, and other boreal [northern] meadow and tundra plants, along with a few twigs, cones, and pollen of high-boreal and tundra trees."[11]

Pfizenmayer also included a brief list of some of the plants found in the mouth and stomach of the mammoth, along with the following comment:

"Thorough investigation established the very interesting fact that the food of the mammoth consisted of the same plants that are still found in the immediate neighbourhood of the discovery, and which we had collected there and preserved for purposes of comparison."[12]

Yes, as widely reported, one of the plant species identified in the lists mentioned above is a species of "buttercup." But the presence of that plant does not necessarily indicate a tropical or sub-tropical or even mild-temperate climate; buttercups grow over a wide range of climate zones, including northern Siberia.

So there is no basis for the conclusion that the Beresovka mammoth had been living in a tropical or sub-tropical climate at the time of its death. On the basis of the food plants found in its mouth and stomach, it had lived in the region where its frozen carcass was found, in northern Siberia.

Frozen condition

The second claim, that the mammoth's body was frozen very rapidly, and inferring catastrophic events, is a mistaken inference from one entry in the daily journal kept by O.F. Herz, the leader of the expedition to recover the mammoth's carcass. Herz wrote:

"In the afternoon we removed the left shoulder ... The flesh from under the shoulder, which is fibrous and marbled with fat, is dark red in color and looks as fresh as well-frozen beef or horse meat."[13]

That quotation is repeated in some publications that promote the view that Earth is a recent creation, including *The Creation-Evolution Controversy*:

"These mammoths, the most notable being the Beresovka mammoth of Siberia, give every evidence of being suddenly buried in mucky water and quickly frozen. (Some estimate temperatures must have been far below minus 100 degrees Fahrenheit)"[14]

And in "The Pre-flood Greenhouse Effect" in *Symposium on Creation II*:

"Birds-eye frozen food experts, in examining the mammoth tissue, have deduced that they were 'thrown in the cooler' suddenly, into temperatures below minus 150 degrees Fahrenheit."[15] [No source is provided for this claim of expert testimony.]

But, if truth be known, that claim of sudden freezing is not consistent with the evidence. The fresh-looking appearance of the mammoth's frozen flesh is deceiving; the mammoth had already undergone extensive decay and putrefaction, probably over a period of days or weeks, before it was completely frozen.

The conclusion that the mammoth had undergone extensive decay before being frozen is supported by several of the daily entries in the journal of Otto F. Herz, the chief of the expedition to retrieve the mammoth, by the testimony of Pfizenmayer, Herz's assistant, in his account of the expedition, and by Tolmachoff a bit later. We will make use of the testimony of these first-hand witnesses to evaluate the claim of "rapid freezing:"

Herz wrote:

"Upon the left hind leg I also discovered portions of decayed flesh, in which the muscular bundles were easily discernible. The stench emitted by this extremity was unbearable, so that it was necessary to stop work every minute."[16]

And:

"Despite the fact that the mammoth is in frozen condition, the stench emitted is very disagreeable."[17]

In anticipation of the severe cold weather and snow that would be coming soon, Herz noted in his journal on September 28 that he would soon order cutting and planing of timber to build a shelter over the mammoth so that they could heat the interior, and continue dismembering the mammoth's carcass after the cold weather arrived. On October 11 he noted that the roof over the shelter had been completed, and on October 13 he reported that the attempt to heat the interior of that structure was successful.

Then:

"After removing the last layer of earth from the back, the remains of food in the stomach were exposed. The latter [the stomach] was badly decayed. We could not continue our work here owing to the solidly frozen condition of everything. After dinner we removed the right side of the abdomen in order to permit the access of heat from the fireplace into the interior of the body."[18]

And:

"We ... cleaned part of the stomach, which contained an immense quantity of food remnants. The walls of the stomach first exposed were dark coffee-brown, almost black in color, and were badly decayed and torn, even where they were not injured mechanically."[19]

And, again:

"Those parts of the stomach that were exposed to the air for any length of time tear even when most cautiously touched, exactly like the membrane beneath the ribs."[20]

According to the testimony of the local hunters who had first discovered the mammoth, its head had apparently been exposed at the surface by a soil slump from the frozen cliff along the Beresovka River in 1900, and parts of the back were exposed by the summer of 1901. Very likely some decay of the animal's flesh had occurred during the summers of 1900-01. For the parts that remained encased in permafrost until excavated by Herz, however, no decay could have taken place since the time that the animal was first frozen and encased in permafrost, those many years ago. The advanced state of decay of the stomach, and of the lower parts of the carcass, therefore, must have occurred after the animal died, and before it was frozen in permafrost.

The account by Pfizenmayer further supports the conclusion that much decay had occurred before the animal was initially frozen, in these words:

> "The well-preserved flesh on the upper parts of the foreleg and thigh and on the pelvis was streaked with thick layers of fat. As long as it was frozen it had a quite fresh and healthy appearance and a dark-red colour like that of frozen reindeer or horseflesh, but it was considerably coarser in fibre. As soon as it thawed, however, it entirely changed its appearance. It became flabby and grey, and gave off a repulsive ammoniacal stench that pervaded everything."[21]

In a section of his paper with the heading "Conditions of Preservation of Frozen Carcasses of the Mammoth and Rhinoceros," Tolmachoff discussed the slow decay and putrefaction of animal carcasses in Arctic climates at length. He wrote:

> "… a strong smell is peculiar to the mammoth localities and to the ground within which remnants are buried, even when they are concealed within and, presumably, still firmly frozen. No process of decay is possible under temperatures below the freezing point, and in the case of the mammoth, rhinoceros, etc., it did not take place …. The smell in the ground may, therefore, be the result of the putrefaction started immediately after the death of an animal, before it became permanently frozen, …"[22]

About stories that were circulated during the time of his writing, Tolmachoff commented:

"... although current opinion attributes to the meat of a mammoth an almost absolute freshness ... As a matter of fact such freshness is a legend. The only proof of it is bright red color of flesh and white or yellowish of fat, and the fact that the flesh used to be devoured with avidity by dogs and wild animals. But the same meat was absolutely unpalatable for an adventurous scientist. All stories published in newspapers of this country [U.S.A.] of a dinner in St. Petersburg where the meat of the Beresovca mammoth was served, are a hundred per cent invention."[23]

Tolmachoff continues,

"An examination of the flesh and fat of the mammoth from Beresovca River has also shown that they suffered a deeply penetrating chemical alteration as a result of the very slow decay which was going on"[24]

and a footnote to this sentence refers to a paper by F.A. Bialinitzki-Birula entitled [in translation] "Histological Observations," p. 19, presumably a paper included in the official *Results of the Imperial Museum expedition* that had been published in Russian.

The frozen mammoth found near the Beresovka River in Siberia died suddenly, to be sure, but almost certainly as a result of individual accident and not by any sudden or widespread catastrophe. The state of decay of parts of the animal that were still embedded in permafrost at the time of the excavation indicates that the freezing took place over the span of days or, more probably weeks, in the cold northern Siberia climate, rather than in minutes.

If truth be known, both the plants found in the stomach of the Beresovka specimen and the partial decay of the flesh before preservation by freezing indicate that this animal had lived in a cold polar climate, and that it died in the far north where its body was found.

It is of some importance that we take note of the dates of publication of the articles and books we have been considering in connection with our discussion of the Beresovka mammoth. The article "The Pre-flood Greenhouse Effect" in *Symposium on Creation II* (1970) was published nearly a decade after the publication of the paper by Farrand (1961), and the book *The Creation-Evolution Controversy* (1976) was published another half-dozen years later. Both were published several decades after the earlier paper by Tolmachoff (1929), the book by Pfizenmayer (1926), and the book by Digby (1926), and still longer after the English translation of excerpts from the report of the excavation of the mammoth by Herz (1904). Of those earlier publications, only Tolmachoff is referred to in a footnote in *The Creation-Evolution Controversy*, and none of them is referred to by footnote or bibliography in "The Pre-flood Greenhouse Effect" in *Symposium on Creation II*. If truth be known, the first-hand observations reported in these earlier publications had already refuted the claims of "tropical or sub-tropical food" in the mouth and stomach of the Beresovka mammoth, and "sudden" freezing of the animal, long before those claims were published in 1970 and 1976.

The claim that the Beresovka mammoth provides evidence supporting the view that the region of the present-day Arctic Circle was once tropical or sub-tropical and that some catastrophe produced a sudden and extreme change in climate is mistaken. If truth be known, there is no support here for the catastrophist view of history.

To its credit, the website www.answersingenesis.org, which promotes the view that Earth is a recent creation, has included the claim that "Woolly mammoths were flash frozen during the Flood catastrophe" in a list of "Arguments that should never be used" in "defending a young-earth creation." On the other hand, the claim continues to be published, as, for example, on the website www.creationscience.com (11/26/09).

You may hold to the view that Earth is a recent creation, and you may hold to a catastrophist view of history for other reasons, if you wish, but I don't think you would want to base your conclusions on incorrect information. Nor would you want your children to be taught incorrect information in school, church, or home. And we certainly wouldn't want stories to be circulated among us Christians without checking whether or not they are true.

Postscript

As noted above, the Beresovka mammoth was excavated in 1901. Much of the mammoth's body was warmed and thawed to be cut apart, then the parts were set outside the heated shelter to be refrozen. The remains were kept frozen for the trip to St. Petersburg during the cold of winter. Eight days after its arrival at St. Petersburg in February 1902, the entire animal was reassembled in the Museum of the Imperial Academy in a steam-heated room and put on display. The mammoth, undoubtedly completely thawed in the warm room, reeked with the stench of decayed and decaying flesh.

The display, apparently, was up for only a short time. Within a few months, the flesh had been stripped from the bones, and the skeleton had been repaired and put on display in the Museum, and it is still there today. During those same few months, about one-third of the original skin and hair of the animal was used in a restoration of the mammoth, in the pose in which it had been found, and the stuffed animal was put on display in the Museum, and it is still there today. Pfizenmayer, who was working at the Museum during that period of time, reported:

> "Both exhibits, together with the most important soft parts preserved in alcohol, are the pride of the St. Petersburg Museum."[25]

The company that first marketed frozen food under the "Birds Eye" brand was established in the early 1960's, according to the history of the company available online. It would be very interesting, I think, to explore when, where, and under what conditions the

"Birds-eye frozen food experts" had examined the mammoth tissue, and deduced that it had been 'thrown in the cooler' suddenly, since the flesh of the mammoth had been put in alcohol to preserve it in 1902.

References

[1] O.F. Herz, [in Russian; title given here in English translation], "Report of the expedition of the Imperial Academy of Sciences to the river Berezovka for excavation of frozen mammoth," *Bulletin of the Imperial Academy* 16, No. 4 (April 1902).

[2] O.F. Herz, "Excerpts from Herz (1902) in English translation," *Report of the Board of Regents of U.S. National Museum, Smithsonian Institution for the year ending June 1903.* (U.S. Government Printing Office, 1904), 611-25.

[3] B. Digby, *The Mammoth and Mammoth-hunting in Northeast Siberia.* (New York: Appleton, 1926).

[4] E.W. Pfizenmayer, *Siberian Man and Mammoth*, translated from the 1926 German edition by Muriel D. Simpson (London: Blackie & Son, Ltd., 1939).

[5] R.L. Wysong, *The Creation-Evolution Controversy*, (Midland, MI: Inquiry Press, 1976), 391.

[6] Donald W. Patten, "The Pre-flood Greenhouse Effect" in *Symposium on Creation II.* (Grand Rapids: Baker Book House, 1970), 21.

[7] V.N. Sukachev, [in Russian; title given here in English translation] "Examination of plant remnants found within the food of the mammoth discovered on the Beresovka River Territory of Yakutsk," in *Results of the Imperial Academy Expedition*, Vol. III, *Scientific Results*, (St. Petersburg: Imperial Academy, 1914).

[8] I.P. Tolmachoff, "The Carcasses of the Mammoth and Rhinoceros Found in the Frozen Ground of Siberia," *Transactions of the American Philosophical Society* 23, (1929), 48.

[9] Henry Fairfield Osborn, *Proboscidea: a monograph of the discovery, evolution, migration and extinction of the mastodonts and elephants of the*

world, (New York: American Museum of Natural History, 1942), 1127.

[10]L.A. Kupriyanova, with B.A. Tikhomirov, "Analysis of pollen from the vegetable remains of food found in the stomach of the Berezovka mammoth," *Proceedings of the Academy of Science U.S.S.R.* 95, No. 6 (1954).

[11]William R. Farrand, "Frozen Mammoths and Modern Geology," *Science* 133 (17 March 1961), 730.

[12]Pfizenmayer, *Siberian*, 101.

[13]Herz, "Excerpts," 621 (diary entry dated October 18, 1901).

> NOTE: The entries in Herz's diary and in the translated entries in Ref. 2, above, are dated according to the "old style" Julian calendar, which was in use in Russia at the time. The dates in this book, like the dates in Ref. 3, above, have been converted to the "new style" Gregorian calendar, by adding 13 days.

[14]Wysong, *Controversy*, 391.

[15]Patten, *Symposium*, 21.

[16]Herz, "Excerpts," 613 (September 24).

[17]Herz, "Excerpts," 618 (October 2).

[18]Herz, "Excerpts," 620 (October 16).

[19]Herz, "Excerpts," 620 (October 17).

[20]Herz, "Excerpts," 623 (October 20).

[21]Pfizenmayer, *Siberian*, 103.

[22]Tolmachoff, "Carcasses," 60.

[23]Tolmachoff, "Carcasses," 60.

[24]Tolmachoff, "Carcasses," 60.

[25]Pfizenmayer, *Siberian*, 145.

3 Brontosaurus Stuff (in Two Parts)

This chapter is in two parts, because it contains two stories. Both of the stories are about the giant plant-eating dinosaur known as *Brontosaurus*, quite likely the dinosaur most familiar to children and the general public. The stories are separate and unrelated to each other. These two separate stories, however, have often been confused and improperly blended together. That improper blend has been (mis)used to aim darts of criticism and ridicule at the science of paleontology in some of the publications that promote the view that Earth is a recent creation.

For example, entering the name *Brontosaurus* in a topic search on the website www.answersingenesis.org brings up several articles. One of these, a reprint of an article entitled "Thunder lizards" from *Creation Magazine*, contains the statement:

"The best known of all the sauropods is *Brontosaurus*, even though 'Bronty' never existed!"

Another entry at that website is a reprint of another article from *Creation Magazine*, entitled "Messages on stone." A footnote to that article states:

"Brontosaurus is of course a misnomer, since it was the result of the wrong head being placed on the wrong body."

A radio preacher told the stories in a similar mixed-up way, and concluded with the comment:

"The making of the brontosaur was like taking the bones of a poodle, topping them off with the skull of a cat, and calling the mismatched skeleton a newly discovered species. ... Most of us have enormous confidence in science, and we can hardly believe that a fossil expert could make a wrong-headed dinosaur."

If truth be known, however, there are two quite separate stories to be told, and each of them, in its own way, provides valuable insights into how the science of paleontology works. So, here are the stories.

STORY #1: WHAT'S IN A NAME?

We humans like to have names attached to things. We name our babies, our trees and flowers, our pets, even our boats and travel trailers. We attach names to various parts of our anatomy, and we name the stars.

Names serve a variety of purposes. Names help us to identify things, and to distinguish one thing from another. Especially, we use names to classify things. But remember, the thing exists apart from its name, or, to put it differently, the name is not the thing.

We use names to classify plants and animals. Linnaeus (1707-1778) developed the system currently in use, using Latin or Latinized names (at a time when people with formal education were taught to read and write that language). We have been naming plants and animals since God gave that mandate through Adam, and new names keep on being applied to newly discovered species of living plants and animals as well as to newly discovered fossil species. We humans go by the genus and species name of *Homo sapiens sapiens*, and the common bullfrog is *Rana pipiens*, etc. Sometimes the discoverer of the new species chooses the name to give honor and credit to him-/herself, as in the fossil genus *Spriggina*, discovered in 1946 by Reginald C. Sprigg in the Ediacaran Formation in Australia. Sometimes it indicates where the species has been found, as in the fossil brachiopod *Spirifer pellaensis*, found in the Pella beds near Pella, Iowa. Sometimes the name indicates some characteristic of the genus or species, as in the fossil colonial coral *Hexagonaria percarinata*, commonly known as "Petoskey stone," in which the part of the skeleton surrounding each individual coral animal is often hexagonal in shape.

31

There is an *International Code of Zoological Nomenclature* that specifies the agreed-upon procedures to be followed in naming a new genus or species. There is an "International Commission on Zoological Nomenclature" that has oversight of the naming of animals, both living and fossil. The Commission is responsible for settling any disputes about names as new species are added to the list, or when old names need to be changed as past errors are corrected and new information is added. The Commission is not called upon to pass judgment on the name of every new species, but responds to appeals that are submitted to it.

Is it *Brontosaurus* or is it *Apatosaurus*?

In the middle 1800's the western parts of North America were being explored by geologists, zoologists, and surveyors, and many new species of plants, animals, and fossils were being discovered. There was often competition among zoologists (for living species) and among paleontologists (for fossil species) to see who would discover the largest number and most spectacular of these new species; the discoverer has the privilege of assigning the name and getting the fame. The rivalry between Edward D. Cope of Philadelphia and Othniel C. Marsh of Yale College was particularly keen, and reports of new discoveries were often rushed into print in order to establish priority.[1] Frequently, in the rush to publication, a name and very brief description were based on only fragments of a specimen, which often resulted in the assignment of more than one name to what turned out to be just one particular genus or species.

There is an interesting history to the naming of the giant plant-eating dinosaur commonly called *Brontosaurus*. It was called a brontosaur in the television series *The Flintstones*, and also on a U.S. Postal Service 25-cent stamp issued in 1989. It is named "Brontosaurus" in many stories, poems, and books, including children's books. *Brontosaurus* is one of the best known dinosaurs in the minds of the general public.

The name for that genus that is recognized and used most frequently in the scientific literature, however, is *Apatosaurus*. Here's the story of how that happened.

In March of 1877 some huge bones were discovered near Morrison, Colorado, by Arthur Lakes. He sent a letter and some specimens to Othniel C. Marsh, asking for identification and some expression of interest in further excavation. Marsh was slow in responding, so Lakes also sent some specimens to Edward D. Cope. Marsh undoubtedly heard that Cope had received some materials, and he promptly contacted Lakes to ask him to continue excavation and to ship the bones to Yale College's Peabody Museum. Lakes then contacted Cope and told him that the bones that had been shipped to him earlier should be forwarded to Marsh. The bones discovered by Lakes turned out to be a large plant-eating dinosaur of a type that had never been seen before in Europe or North America, and now the competition between Marsh and Cope intensified. Both were excavating for more bones, first at Morrison, Colorado, then near Cañon City, Colorado, and later at Como Bluff, Wyoming.

Lakes excavated bones of two large dinosaurs from the site at Morrison, Colorado in 1877, with some help from Benjamin Mudge and Samuel Williston, both of whom had excavated for Marsh before, and shipped the material to Marsh. Marsh hurried a brief description into print, stating that one of those skeletons was "nearly complete, in excellent preservation," although his published description is less than one page with no illustrations of the bones. He named the specimen *Apatosaurus ajax*. The second skeleton belongs to the same genus, and Marsh named it *Apatosaurus grandis*.[2]

In August of 1877 Marsh received a letter from two men under fictitious names informing him that they had found many large bones near the Laramie Station of the Union Pacific Railroad in Wyoming. Marsh apparently replied and asked them to ship the bones to him. The two men, whose real names were W.E. Carlin and W.H. Reed, sent the shipment in mid-September. Realizing that this was a very important discovery, Marsh sent a telegram to Samuel Williston in

Colorado, urging him to go to Wyoming to supervise further excavation. The find was at Como Bluff, one of the most important localities of fossil dinosaurs in western North America. In his book *Dinosaurs of North America* Marsh wrote that he had "secured from it the remains of several hundred dinosaurs."

In 1879 a large skeleton was excavated from Quarry 10 at Como Bluff, primarily by William H. Reed and E.G. Ashley (although Arthur Lakes had also done some excavating at Quarry 10 earlier). This skeleton, also, was nearly complete, and is now exhibited in the Peabody Museum at Yale. Marsh named it *Brontosaurus excelsus*.[3] It is still one of the finest "brontosaur" skeletons ever found.

In 1903 Elmer Riggs of the Field Museum in Chicago published the description of a dinosaur skeleton that he and H.W. Menke had collected from the Grand River Valley near Fruita, Colorado in 1901. He classified the specimen as *Apatosaurus*. In connection with that description he also undertook a re-study of Marsh's dinosaurs. He found Marsh's *Apatosaurus* and *Brontosaurus* to be very similar, and he thought them to be two specimens of the same genus, with *Apatosaurus* being a more juvenile specimen. In his monthly column in *Natural History* entitled "Bully for Brontosaurus," (reprinted in the collection of the same title) Stephen Gould wrote:

> "No big deal; it happens all the time. Riggs rolled the two genera into one in a single paragraph."[4]

That paragraph from Riggs reads as follows:

> "The genus *Brontosaurus* was based chiefly upon the structure of the scapula and the presence of five vertebrae in the sacrum. After examining the type specimens of these genera, and making a careful study of the unusually well-preserved specimen described in this paper, the writer is convinced that the Apatosaur specimen is merely a young animal of the form represented in the adult by the Brontosaur specimen. ... In view of these facts the two genera may be regarded as synonymous. As the term '*Apatosaurus*' has priority, '*Brontosaurus*' will be regarded as a synonym."[5]

The choice by Riggs of the name *Apatosaurus*, with *Brontosaurus* as a synonym, was done in accord with the *Code* as it existed at the time. O.C. Marsh had died in 1899, and no one appealed to the International Commission for a decision on Riggs' proposal. Other paleontologists apparently have accepted Riggs' conclusion, and the name *Apatosaurus* is now applied to that genus throughout the scientific literature. Though officially *Apatosaurus*, the name is often published with recognition of the synonym, as follows: "*Apatosaurus (=Brontosaurus)*." Either name is correct; *Apatosaurus* is preferred. That's the way the naming system works.

Is it OK to have two names for the same thing? Of course it is. The North American "mountain lion" is also known as "cougar" or as "puma." No problem.

The general public and the popular media - even encyclopedias - have been slow to adopt the name *Apatosaurus* for that genus. While writing this chapter I was chatting with a friend, a college graduate and college teacher, about these names; she knew about *Brontosaurus*, but when I mentioned the name *Apatosaurus* she said, "Never heard of it." The 1960 edition of *World Book* encyclopedia has been in my home since our children were small, and it uses *Brontosaurus* throughout its article on dinosaurs, and does not mention *Apatosaurus*. *Encyclopedia Britannica*, widely respected as the quality standard of the industry, uses "*Brontosaurus (Apatosaurus)*" to introduce the genus to its readers in its fourteenth edition, printed in 1970, and then uses *Brontosaurus* throughout the remainder of its article on dinosaurs. Change (progress?) does occur, however; the 2002 fifteenth edition of *Macropedia Britannica* introduces the genus as "*Apatosaurus* (formerly known as *Brontosaurus*)," and then uses *Apatosaurus* throughout the remainder of its entry on dinosaurs. Perhaps some day the general public will get used to the name *Apatosaurus*, and will have forgotten all about *Brontosaurus*. Or maybe not. No matter; it's the same creature, either way. The name is not the thing.

Paleontology works. Slowly, sometimes, but it works.

STORY #2: THE HEADLESS DINOSAUR

Several skeletons of giant plant-eating dinosaurs were found in North America and elsewhere in the world during the late 1800's, and new discoveries of similar creatures continue to be made from time to time. There are many similarities among those fossils, but also sufficient differences to classify them as separate genera (plural of genus). Those that feature in this story include *Camarasaurus*, *Diplodocus*, and, of course, *Apatosaurus* (=*Brontosaurus*). Sometimes the skeletons of these different genera were found in separate locations; sometimes those that lived during the same period of history were found together.

By 1877 some specimens of *Camarasaurus* and of *Diplodocus* had been found with skulls articulated with the neck bones, in the same positions as they were in life. However, no skull was found with the *Apatosaurus* (=*Brontosaurus*) skeleton that was excavated for Marsh by Reed and Ashley from Quarry 10 at Como Bluff, Wyoming in 1879. Marsh wanted to publish a description of the skeleton from Como Bluff, and preparations were being made to exhibit the skeleton at the Yale Peabody Museum. Should a description be published without a skull? Should the skeleton be mounted in the Museum without a skull?

Presumably on the basis of similarities in other parts of the skeletons, Marsh decided to use *Camarasaurus*-like partial skulls that had been found at Quarry 13 East at Como Bluff for his reconstruction of the specimen that he had named *Brontosaurus*, although Quarry 13 East is a few miles from Quarry 10 where the skeleton without its head had been excavated. So the reconstruction by Marsh was published with a *Camarasaurus*-like skull.[6]

The pelvis and hind limbs of the specimen from Como Bluff Quarry 10 were mounted and put on display in the old building of the Yale Peabody Museum in 1901. That display was dismantled around 1914, prior to the razing of the old building. In the late 1920's the entire skeleton was erected in the new Museum building, and, in accordance with the reconstruction published by Marsh in 1883, a *Camarasaurus* skull from the Felch Quarry at Garden Park, Colorado, was placed on the *Apatosaurus* (=*Brontosaurus*) skeleton for that display.

Unfortunately, the other available *Apatasaurus* (=*Brontosaurus*) skeletons also lacked associated skulls. The two specimens recovered for Marsh from Morrison, Colorado in 1877, which he had named *Apatosaurus*, had also been found without a skull articulated with the neck vertebrae. The specimen of *Apatosaurus* (=*Brontosaurus*) that was excavated from Fruita, Colorado by Elmer S. Riggs for the Field Museum in Chicago in 1901, and set up for display in 1903, was also a partial specimen, with the skull and most of the neck bones missing, eroded away before the specimen was discovered and excavated. So there was still no skull that was unambiguously associated with *Apatosaurus* (=*Brontosaurus*).

Meanwhile, in 1909, Earl Douglass excavated a nearly complete skeleton of *Apatosaurus* (=*Brontosaurus*) from a quarry near Jensen, Utah for the Carnegie Museum in Pittsburgh. (That quarry has produced numerous additional dinosaur specimens, and has now been incorporated into Dinosaur National Monument.) The Carnegie Museum skeleton included all of the neck vertebrae, including the atlas, the vertebra to which the skull would have been attached during life. There was no skull articulated with the neck vertebrae, but there was a skull (without the lower jaw) found about twelve feet from the atlas of that specimen. The skull and the remainder of the *Apatosaurus* (=*Brontosaurus*) skeleton were shipped to the Carnegie Museum where the bones were removed from the rock, and the skull was found to fit exactly into the atlas of the neck of that skeleton. However, it could not be claimed with certainty that the skull belonged to the *Apatosaurus* (=*Brontosaurus*) skeleton, since it had been found twelve feet from the neck vertebrae.

The skull found by Douglass at Dinosaur National Monument was very similar to the skull of *Diplodocus*, and very unlike the skull of *Camarasaurus* that Marsh had used for his proposed reconstruction of the skeleton he had named *Brontosaurus*. Marsh was held in high regard by other paleontologists, so when the skeleton was mounted for display at Carnegie Museum, W.J. Holland, then Director of the Museum, was persuaded not to express public disagreement with Marsh's proposed reconstruction by placing the *Diplodocus*-like skull on the mounted display. However, Holland refused to use what he thought was the wrong skull, so the skeleton was placed on display in the Museum in 1913 without a head. It remained headless until 1932.

After W.J. Holland and Earl Douglass had died, a *Camarasaurus*-like skull was placed on the *Apatosaurus* (=*Brontosaurus*) skeleton at Carnegie Museum. That choice was defended in a paper published by C.W. Gilmore, who claimed that the skull which Holland had referred to as probably belonging to the *Apatosaurus* (=*Brontosaurus*) had actually been found more than one hundred feet from the skeleton, and not twelve feet, as Holland had claimed.[7] (Gilmore based that claim on the testimony of one of the assistants who had aided Earl Douglass in the excavation. See the next paragraph for more on this topic.) Although Gilmore was instrumental in placing the *Camarasaurus*-like skull on the Carnegie *Apatosaurus* (=*Brontosaurus*) skeleton in 1932, he stated in his 1936 paper that:

> "After a review of all of the various skull parts that have been attributed to the genus *Apatosaurus*, it becomes quite evident that not in a single instance has there been such an association that we can definitely say this is a skull or a portion of a skull of *Apatosaurus*. At the present time we must therefore consider the skull of *Apatosaurus* as unknown."[8]

Thus, despite the acknowledgment of uncertainty, the *Camarasaurus*-like skull stayed on the skeleton at Carnegie Museum until 1979.

In 1975 David S. Berman, Assistant Curator, Section of Vertebrate Fossils at Carnegie Museum and John S. McIntosh, Wesleyan University, Middletown, Connecticut undertook a detailed review of the entire history of the skeletons of *Apatosaurus* (=*Brontosaurus*) and other giant plant-eating dinosaurs. Berman and McIntosh identified numerous errors that had been made in earlier reconstructions and in mounting skeletons in several museums, errors which had (mis)led various people to conclude that there were more similarities between the skeletons of *Apatosaurus* (=*Brontosaurus*) and *Camarasaurus* than is actually the case. They were also able to show many similarities between *Apatosaurus* (=*Brontosaurus*) and *Diplodocus* that had not been recognized before. They concluded that O.C. Marsh and C.W. Gilmore had been mistaken in choosing a *Camarasaurus*-like skull for *Apatosaurus* (=*Brontosaurus*), and that W.J. Holland had been correct in thinking that he and Earl Douglass had found the skull belonging to *Apatosaurus* (=*Brontosaurus*) about twelve feet from the atlas of the neck. (The assistant to Earl Douglass had apparently confused one skull with another in the testimony that Gilmore had reported in his 1936 paper.) The study by Berman and McIntosh reporting the evidence they had accumulated was published in the *Bulletin of Carnegie Museum of Natural History.*[9] The officials at Carnegie Museum became convinced that Berman and McIntosh were correct, and a "beheading" ceremony was held at the Museum in 1979, removing the *Camarasaurus*-like skull from the *Apatosaurus* (=*Brontosaurus*) skeleton and replacing it with a *Diplodocus*-like skull modeled after the skull found at Dinosaur Quarry by Earl Douglass in 1909.

In 1982 the officials at Yale Peabody Museum held a similar ceremony, replacing the *Camarasaurus* skull on their specimen of *Apatosaurus* (=*Brontosaurus*) with a *Diplodocus*-like skull.

Hopefully, this isn't the end of the story. To this date, no *Apatosaurus* skeleton has been found with a skull unquestionably belonging to the skeleton. There is still some small question as to whether the skull found twelve feet from the atlas of the skeleton excavated by Douglass in 1909 actually had been attached to the rest of the skeleton when that creature was alive. So we are still not entirely certain that the skull that has been chosen as belonging to *Apatosaurus* (=*Brontosaurus*) is, in fact, the correct one. Perhaps you, dear reader, will be the person who some day, somewhere, finds an *Apatosaurus* (=*Brontosaurus*) skeleton with its skull articulated with the neck bones of the specimen; you will become famous, and you will deserve the gratitude of many paleontologists. Perhaps the find will be the occasion for some more beheadings of *Apatosaurus* (=*Brontosaurus*) skeletons in museums around the world; perhaps not.

So, those are the stories, told one by one.

Misinterpretations

As noted at the beginning of this chapter, there are some publications that promote the view that Earth is a recent creation in which these events of naming new species and attempting to identify the proper skull for *Apatosaurus* (=*Brontosaurus*) are confused and improperly merged, and then held up as an example of failure in science. For example, an article (a "blog") that was posted on the website <angelfire> led off with the headline "There is no such thing as a Brontosaurus" and the article claims, mistakenly, that O.C. Marsh "purposely" placed the wrong head on a skeleton of an *Apatosaurus* and then claimed that it was a new species, *Brontosaurus*. Actually, if truth be known, the names of *Apatosaurus* and *Brontosaurus* had been applied to different headless skeletons well before any suggestion was made as to which skull belonged to that genus.

In hindsight, it seems obvious that Marsh was hasty in giving the name *Brontosaurus* to the specimen from Como Bluff when he already had bones of *Apatosaurus* (from Colorado) in his possession. We note, however that the naming system works. Sometimes there is duplication, and a re-study finds that two specimens with different names are actually the same genus. No big deal; the correction is published, and we go on from there.

There are many areas in science in which we have incomplete information. Without a definitive specimen for the proper skull of a fossil skeleton, a best-guess choice is sometimes made; when better information is available, a correction is made. Using whatever information we have available at the time, scientific explanations are formulated by imperfect humans.

If truth be known, the two separate stories are very nice examples of how science works, and are a demonstration of the openness of science to admitting and correcting mistakes, and the openness of science to incorporating reliable new information into its explanations and claims. The process is sometimes slower than we might wish, but it works. Those events in the history of our study of *Apatosaurus* (=*Brontosaurus*) do not in any way justify heaping ridicule on science or scientists. Nor do they justify the mis-reporting of the history of these controversies.

References

[1] Url Lanham, *The Bone Hunters*, (New York and London: Columbia University Press, 1973).

[2] Othniel C. Marsh, "Notice of New Dinosaurian Reptiles from the Jurassic formation," *American Journal of Science* 14, series 3 (1877), 515.

[3] Marsh, "Notice of New Jurassic Reptiles," *American Journal of Science* 18, series 3 (1879), 503.

[4] Stephen Jay Gould, *Bully for Brontosaurus*, (New York and London: W.W. Norton & Company, 1991), 89.

[5] Elmer S. Riggs, *Structure and Relationships of Opisthocoelian Dinosaurs: Part I. Apatosaurus* Marsh. (Chicago: Geological Series, Vol. II, No. 4. Publication 82, Field Columbian Museum, 1903), 170.

[6] Marsh, "Principal characters of American Jurassic dinosaurs, Part VI: Restoration of Brontosaurus." *American Journal of Science* 26 (1883), 81-86 and Plate 1 (following 496).

[7] Charles W. Gilmore, "Osteology of *Apatosaurus*, with special reference to specimens in the Carnegie Museum," *Memoirs of the Carnegie Museum*, XI, No. 4 (1936), 177-281.

[8] Gilmore, "Osteology," 190.

[9] David S. Berman and John S. McIntosh, "Skull and relationships of

the upper Jurassic sauropod *Apatosaurus*," *Bulletin of Carnegie Museum of Natural History*, No. 8 (1978), 30.

4 A Whale and a Tale

In April 1976 the front-loader operator at a quarry near Lompoc, California uncovered some fossil bone. Dr. Lawrence Barnes, vertebrate paleontologist at the Natural History Museum of Los Angeles County ("LACM"), was contacted, and scientists from the museum proceeded to excavate the skeleton of a large baleen whale, about 80 feet in total length.

The whale skeleton was found in a layer of "diatomaceous earth," a deposit consisting mostly of the skeletons of diatoms, single-celled algae that live in ocean waters. When they die, their tiny silica skeletons fall to the ocean floor, commonly deposited with clay to form diatomaceous earth. Diatomaceous earth is used as a filtration aid, as a mild abrasive, and as an absorber for liquids (as in cat litter). This deposit was being mined by Grefco, Inc. at the location known as the Miguelito Mine (also known as the Tolbert lease).

The story of the discovery of the fossil whale and its excavation by LACM scientists was reported in a news item published in the Santa Barbara News Press. The accompanying photo showed several workers chipping away the rock surrounding the skeleton, working on the steep slope of the quarry wall. The whale was found lying at an angle of about 50 degrees from horizontal, with its skull up and its tail down.

The news was picked up and reported in the scientific publication *Chemical and Engineering News*, on the "Newscripts" page of the October 11, 1976 issue.[1] That news item included the statement:

"The whale is standing on end in the quarry and is being exposed gradually as the diatomite is mined."

The story was also reported in *Chemical Week*.[2] Neither of those brief reports mentioned anything about the rock layers adjacent to the diatomaceous earth in which the whale skeleton had been found, or the orientation in space of the layer in which the whale skeleton was embedded.

The proponents of the idea that Earth is a recent creation, and of the idea that much or most fossil-bearing rocks on Earth resulted from a worldwide flood event, responded promptly to the article. The "Letters" column in *Chemical and Engineering News* of January 24, 1977 published a letter commenting on the whale skeleton as follows:

"K.M. Reese made no comment concerning the implications of the unique discovery of a baleen whale skeleton in a vertical orientation in a diatomaceous earth quarry in Lompoc, Calif. However, the fact that the whale is standing on end as well as the fact that it is buried in diatomaceous earth would strongly suggest that it was buried under very unusual and rapid catastrophic conditions. The vertical orientation of the whale is also reminiscent of observations of vertical tree trunks extending through several successive coal seams. Such phenomena cannot easily be explained by uniformitarian theories, but fit readily into an historical framework based upon the recent and dynamic universal flood described in Genesis, chapters 6-9."[3]

An entry by the same author, containing much the same content, was published in the "Panorama of Science" section of the *Creation Research Society Quarterly* in June 1977.[4]

The story of the "whale on its tail" was retold in the book *The Controversy: Roots of the Creation-Evolution Conflict*. In a section headed "Geological Evidences," the news article in *Chemical and Engineering News*, along with the letters that followed, was reprinted, along with the comments:

"The evidence that pointed to catastrophism has not disappeared. It is still the same today. In fact, even more evidence of catastrophism is being continually discovered."[5]
"Evidence of catastrophe is clearly present. The atmospheric disturbances that must have taken place to cause such wholesale destruction are difficult for us to even imagine. They would fit nicely,

however, with the events described in Genesis 7 and 8."[6]

If truth be known, however, the whale skeleton actually lay parallel to the bedding layers in the deposit of diatomaceous earth. Both the whale skeleton and the diatom deposits were laid down in horizontal layers, and the entire series of rock layers, including those above and below the diatomite, have been tilted to their present orientation by tectonic forces after the sediments had been deposited. Thus, the whale skeleton was tilted along with the rock layer in which it was buried.

The claim of a whale "buried on its tail" was based on a misunderstanding. Possibly, the claim was a mistaken inference from one sentence in the *Chemical and Engineering News* article:

"The whale is standing on end in the quarry and is being exposed gradually as the diatomite is mined."

That was a correct factual report, but did not tell the whole story. The mistaken inference simply assumed that the whale skeleton was "vertical" while the bedding of the surrounding sediments was horizontal, based on partial information. Finding out "the rest of the story" would not have been difficult; the photo that was published in the *Santa Barbara News Press* along with the story had clearly shown the tilted rock layers exposed in the diatomite quarry.

Prior to 1995, there is no indication in the publications that promote the view that Earth is a recent creation that the context of the find was ever investigated; consequently, an incorrect conclusion was published and continued to be spread in the Christian community. As late as 1988, a brief anonymous story on the "Our World" page in the magazine *Creation ex Nihilo* mistakenly listed the whale found in the diatomite quarry at Lompoc, California as an example of a "polystrate" fossil, that is, a fossil that cuts across the layering in the sedimentary deposit in which it is found.[7]

The original publication and later promotion of this sort of mistaken claim for evidence supporting catastrophism and flood geology could and should have been avoided. Surely, this sort of "rookie" mistake is rare, you would think. Sad to say, claims of scientific support for the view that Earth is a recent creation are all too often based on statements from the scientific literature taken out of context, or taken alone without consideration of additional important information about the topic being considered. More examples will appear in later chapters of this book.

It should be added that (finally) in 1995 an article was published in the *Creation ex Nihilo Technical Journal* stating that:

"Contrary to some reports that have circulated, the 80-90 ft (24-27 m) long fossilized baleen whale found in April 1976 in an inclined position in a diatomite unit in the Miguelito Mine at Lompoc, California, was not buried while 'standing on its tail.' An onsite investigation has revealed that the diatomite unit which entombed the whale is also inclined at the same angle, the whale having been buried in the diatomite unit while both were in the horizontal position, and subsequent earth movements having tilted both."[8]

In the intervening time, a mistaken claim had been published and widely distributed in the Christian community.

References

[1]K.M. Reese, "Workers find whale in diatomaceous earth quarry," *Chemical and Engineering News* 54, No. 42 (October 11, 1976), 40.

[2]Anonymous, "Behemoth found in quarry," *Chemical Week* 119 (October 13, 1976), 15.

[3]Larry S. Helmick, "Strange Phenomena," *Chemical and Engineering News* 58, No. 4 (January 24, 1977), 5.

[4]Helmick, "Whale Skeleton Found in Diatomaceous Earth Quarry," *Creation Research Society Quarterly* 14, No. 1 (1977), 71.

[5]Donald E. Chittick, 1984. *The Controversy: Roots of the Creation-Evolution Conflict.* (Portland: Multnomah Press, 1984), 218.

[6] Chittick, *Controversy*, 221.

[7]Anonymous, "Polystratic Fossils," *Creation ex Nihilo* 10, No. 2 (1988) 25.

[8]Andrew A. Snelling, "The Whale Fossil in Diatomite, Lompoc, California," *Creation ex Nihilo Technical Journal* 9, No. 2 (1995), 244.

5 History in the Rocks

All of us have a history. We were born, we grew (or are growing) up, etc., etc. Even when we have become old, if our minds are still healthy, we can remember events and experiences of our early years. Those memories are apparently imbedded in the cells of our brains.

Our nation has a history, too, whichever nation in the world is being referred to as "our nation." Some nations in today's world are relatively young, and some have existed as a nation for hundreds or thousands of years. For many nations, however, the birth of the nation is not within the memory of any of its living citizens. The most important evidence from which to learn about the history of our nation consists of written documents. We even have written documents providing evidence for the histories of many nations and tribes that occupied our planet for a time during the past several thousand years, although they no longer exist as distinct nations today.

Trying to learn the histories of peoples and tribes and nations that have not left any written records presents a somewhat different challenge. We still can get some information about them, however, from examples of their art, the architecture of their buildings, and remnants of material objects of their living habits which have been preserved to the present day. Archeologists in virtually every nation on Earth are engaged in excavations of evidence of the earlier history of the peoples who at one time occupied the region. Our understanding of the history of some people or nation based on written documents is certainly more detailed, and perhaps more accurate, than a history based only on non-written artifacts, but we can learn a good deal of history by inference from those artifacts. If the archaeologist/historian is honest, he/she will be appropriately modest and tentative in publishing her/his interpretation of the data, because there may well be several interpretations that are equally likely on the basis of the available evidence.

Rocks, also, have a history. That history is certainly not written in human language. There are clues, however, from which we can infer - perhaps a better term would be "reconstruct" - the history of a rock or rock structure. The clues are in the rocks themselves. So what sorts of clues should we look for, and how do we go about reaching a reasonable interpretation of what we find? (This will be a condensed version; I am not going to give you an entire introductory geology course.)

We start with the three main categories of rock types, known to most of us from middle-school science, or from the "20 Questions" game: igneous, sedimentary, and metamorphic. Those are categories of the mode of formation of the rock. 1) Igneous rocks are formed from hot, molten material that has come from deep within the Earth and has cooled and solidified at or near Earth's surface. Granite and volcanic lava are examples of igneous rocks. 2) Sedimentary rocks are of two sub-types. One type is called "clastic" sedimentary rock because it is made up of "clasts," or fragments, of other rocks that have been broken apart by weathering, and those clasts have been cemented (glued) together by some natural cementing agent. Sandstone and shale are examples of clastic sedimentary rocks. The other type is called "chemical" sedimentary rock, and the particles that make up the rock have been deposited from water solution, either by some chemical process or by evaporation of the water; cementation of the particles may be a part of the process in some cases. Limestone and rock salt are examples of chemical sedimentary rocks. 3) Metamorphic rocks are formed when previously existing rocks are subjected to fairly intense heat and pressure, but did not get hot enough to melt. The heat has made it possible for the atoms in the rocks to move, perhaps forming new minerals, and often with evidence of alignment of some minerals to form parallel lines or flat surfaces resulting from the pressure that has been exerted on the rock by forces within Earth's crust. Slate and marble are examples of metamorphic rocks.

Recognizing the characteristics that allow us to classify a rock into one of those three categories has already told us something about the way in which the rock was formed.

In sedimentary rocks, especially, there are often some structures that tell us something about the surrounding conditions in which the rock was formed. We may find ripple marks in the rock, especially common in sandstone and shale, which tell us something: if the ripples are symmetric, with the angle of the slope on both sides of the crest of the ripple being the same, we can confidently infer that the particles were deposited in shallow, standing water by the action of waves on the water surface. We make that inference because that is where we find symmetric ripple marks along shorelines today. If the ripple marks are asymmetric, with the slope on one side of the crest being steeper and the other side being more shallow, we can infer that the particles were deposited by running water, as in a stream bed, or by the wind, as in a sand dune. That's where we find asymmetric ripple marks being formed today. By noting which side of the ripple mark is at the steeper angle, we can tell which direction was downstream, or downwind. Often we can distinguish sand grains that were deposited in a running stream from those deposited in the wind because sand grains that have been colliding with each other in the wind have a distinctive surface texture. We can study sand grains in those two different environments today, and learn how to tell the difference, and we apply that knowledge to our study of rocks that were deposited at some time in the past. Sometimes we find mud crack patterns in sedimentary rocks, with exactly the same kind of pattern that we see in dried up puddles or ponds today, and we infer that those rocks were formed by particles that were in an environment where they were covered with water part of the time, and dried up part of the time, probably as the local seasons changed.

There are many other examples that could be given, but those few illustrate the kind of approach that is commonly used to learn about the history of rocks. We study present-day processes to find out what sorts of patterns we can note in the results; then we infer similar processes of the past where we note the same sorts of patterns in rocks that were formed in the past. This approach has been given a name; it is called "uniformitarianism," or more formally, "The Principle of Uniformity."

We certainly are uniformitarian in our everyday approach to learning about events of the recent past. If we come home from an evening out or a short trip out of town, find the front door ajar and the furniture in disarray, we say, "Someone broke into the house while we were gone." If we find the dresser and desk drawers pulled out and contents strewn over the floor, we say, "Whoever was here was looking for something." If we do a quick inventory of our valuables and find some missing, we say, "Whoever was here is a thief." The police dust various surfaces for fingerprints, maybe take an impression of tire tracks, etc. etc. Like any decently written murder mystery, the clues that were left behind are used to infer a likely scenario of what happened in the past. But the clues are observed in the present.

Another example (a favorite of mine) is a history of rabbits and rabbit tracks. We live in Michigan, where it sometimes snows in winter. One morning I got up, looked out the window, and saw some familiar marks in the fresh snow on the yard, and I said, "There was a rabbit in the yard." I had not seen the rabbit that morning, but I have seen cottontail rabbits hopping in the snow before, and the pattern of tracks that they leave behind is distinctive and easily recognizable. Well, I guess you have to be able to distinguish the tracks of a cottontail rabbit from those left by a dog, or an elephant, or a red squirrel, in order to identify the tracks in the snow in my yard that morning as those of a rabbit. But we infer the events of the past on the basis of the clues found at the present. Could I be absolutely, positively certain that the tracks were left by a rabbit? Well, no, they could have been left there, just to fool me, by my practical joker cousin, who had hopped through my yard on a pogo stick that he had built to leave tracks just like those of a rabbit. So there are some other highly remote possibilities, but most likely the tracks were left by a genuine rabbit.

Similarly with the history of rocks. Sandstones with symmetric ripple marks were most likely formed from sand grains deposited in shallow standing water.

The uniformitarian approach to learning about history has been given a bad rap in the publications that promote the view that Earth is a recent creation, but the criticism is undeserved. Part of the problem is an apparent misunderstanding of the way in which the practice is carried out. The book *Scientific Creationism* devotes a chapter of 40 pages to the topic of "Uniformitarianism or Catastrophism?" In that discussion, uniformitarianism is identified as the idea that:

"the fossils and the rocks and the other features of the earth's crust formed slowly over vast aeons of time by the same processes now at work in the earth."[1]

Later in that chapter, the idea of uniformitarianism is referred to as:

"present-day geological processes, acting at the same rates as at present."[2]

But, if truth be known, restricting processes of the past to operation "at the same rates as at present" is certainly not the creed of the modern geologist. It is true that a more rigid notion of a constancy of rates of natural processes in the Earth was defended by Charles Lyell in the early-to-middle 1800's than is held to by geologists of today. Even Lyell was not as rigid about constancy of process rates as is sometimes supposed; everyone can readily see that local erosion rates are varying as local rainfall amounts vary. In a broad way, however, geologists of today continue to think that similar processes acting in the past under similar conditions are likely to produce the same kinds of results at about the same rates as in the present. Geologists of today also continue to accept and use the idea that the processes that formed rocks and rock structures in the past are not different in character from the processes forming similar deposits and similar rock structures that are open to our observation today.

We should note, however, that geologists do not limit themselves to inferring past events on the basis of events and processes of exactly the same sort as are directly observable today. There are many active volcanoes in the world today, and we can observe the behavior of various types of lava produced by molten magma that is forced to Earth's surface from deep within the crust. The lava flows of Hawaii are among the most fluid (lowest viscosity) lavas that are known to be forming at the present time, but they are still quite viscous compared with, say, water. Hawaiian lava will not flow downslope unless the slope is at a steeper angle than about 3 degrees. Yet there are extensive deposits in the Deccan Plateau of India, and in the Columbia Basin of the Pacific Northwest of North America, that consist of rock that is almost certainly produced by volcanic activity. We know that because of various characteristics that are like those of present day volcanics. The viscosity of those lava flows, however, must have been much lower than that of any present-day volcanics. One such lava flow in the Columbia Basin, about 80 feet thick, can be traced over an area of about 20,000 square miles. It must have flowed out with a viscosity close to that of water in order to cover such a large area without congealing and ceasing to flow as it cooled. These lavas, with a chemical composition that identifies them as basalt, are commonly known as "flood basalts," or "flood lavas." Our understanding of the past events that produced those extensive layers of volcanic rocks, then, is an extension beyond any present-day observations.

Geology has always had room within its ideas for catastrophes. Local catastrophic events such as earthquakes, explosive volcanic eruptions, water floods, tsunamis, and many others have been part of human experience throughout human history. There is also abundant evidence that such local catastrophes, perhaps very extensive in their effects, have occurred on Earth in the past, based on our observations of the effects of such catastrophes that are observed at the present time. Such catastrophes are included in modern geological interpretations of the past, based on uniformitarian arguments. Modern interpretations of the geological history of the Earth, however, consider the major fraction of the rocks found on Earth to be the product of less dramatic processes, processes very much like those that are going on from day to day in our present day world.

If you wish to learn more about modern geological interpretation of the history of rocks, I recommend the book *Geology Illustrated*.[3] The first part of the book deals with present day geological processes, and the second part presents numerous case histories in which those geological processes provide the basis for the historical interpretation of some regions of North America, primarily western United States. Of special interest is the chapter on the history of the rocks exposed in the Grand Canyon of Arizona. The book is currently out of print, but you should be able to find a copy in the local public or college library.

Appearance of age

We should add a word about an idea that seems to be attractive to some Christians, namely, that God created the world with the "appearance" of age, although he actually created it recently. According to that idea, he created the rocks with radioactive isotopes and the products of the decay of those radioactive isotopes already in the rocks, so that they appear to be old by radiometric dating, but they really are very recent. He also created distant stars with their light rays already reaching the Earth at the moment of a recent creation, even though, by all our human measurements, they appear to be billions of light years away.

That idea seems to give an advantage to those who believe it, because then they do not need to concern themselves with any of the scientific evidence that might otherwise lead us to think that the world is old, because that evidence isn't real, anyway. It only looks real. And, they say that, after all, God "could have" created the universe in that way, because he is almighty.

Yes, God can do what he chooses to do. He is almighty. There is no way to disprove the suggestion of "appearance" of age in a recent creation.

However, that idea is an unsafe shelter for those who live in it. There are at least two reasons why that idea should be rejected by all of us Christians:

1. Do you think that God is the kind of being who would hop through the snow in my back yard on a pogo stick with feet shaped like the tracks of a cottontail rabbit, just to fool me, like my practical joker cousin might? (Or maybe God wouldn't even need a real pogo stick.) I think that God has a sense of humor; I can imagine that he chuckled to himself as he smuggled Moses into the Pharoah's palace as an adopted member of the family! But I don't think that God would engage in the practical joker, "Let's see if we can fool 'em" sort of prank. If he were that sort of being, how could he inspire the Psalmist to affirm "The heavens are telling the glory of God, the skies proclaim the work of his hands" (Psalm 19) if the tales that the heavens tell, and the proclamations of work done, were only make-believe? Is the glory of God real? We would not be able to affirm that the glory of God is real unless the skies proclaim real work, and the heavens tell true stories. If he were the practical joker sort of being, how could the writer of Romans 1:20 affirm that "God's invisible qualities - his eternal power and divine nature - have been clearly seen, being understood from what has been made?"

2. If the history in the rocks and in the stars is not real history, then there is no real history. After all, God "could have" created the universe just five minutes before you read this sentence. He "could have" created library buildings filled with books whose copyrights span the centuries, with some of the pages turned brown and worn at

the edges from "apparent" use, so that they look old, although they were created only five minutes ago. He "could have" created you (and me) with memories of a childhood that never actually occurred. The pinch comes when we realize that he "could have" created a gospel telling us about a Jesus Christ who never really existed. I wouldn't like that idea, would you?

No, my friend. History is not concerned with what God "could have" done, but with what he actually did. History is real. Jesus Christ was and is real. He really died, and he really was raised from the dead. And if the evidence indicates that the rocks and the distant stars are old, then they really are old.

References

[1] Henry M. Morris, Ed., *Scientific Creationism*, (San Diego: Creation-Life Publishers, 1974), 91.

[2] Morris, *Creationism*, 101.

[3] John S. Shelton, *Geology Illustrated*. (San Francisco and London: W.H. Freeman and Company, 1966).

6 Mt. St. Helens and Catastrophism

The explosive eruption of Mt. St. Helens on May 18, 1980 was a catastrophe, no doubt about it. But it was not even close to being a worldwide catastrophe.

The explosive eruption blew away the top 1700 feet of the mountain, and produced a massive rock slide that moved northward into Spirit Lake and deposited hundreds of feet of rock slide debris on the north edge of the mountain and the valley of the North Toutle River. Volcanic ash was spread over a wide area north and northeast of the mountain. The memory is vivid for many of us.

The suddenness of the eruption provided the occasion for publication of the claim that other geological structures, including the Grand Canyon of Arizona, could be the result of similarly sudden events. For example:

"The explosion uprooted tremendous numbers of trees and blew out vast quantities of soil and rock, transporting them rapidly in giant mud flows downhill into Spirit Lake, near the base of the mountain, and into the basin of the north fork of the Toutle River. Up to 600 feet of *laminated* sediments were deposited in a single day, looking exactly like stratified beds normally interpreted to represent long ages. Furthermore, a subsequent mud flow scoured out a deep canyon through these sediments, in essentially another single day. The resulting canyon section, with its stratified walls, looks much like a scale model of the Grand Canyon. The whole phenomenon constitutes a graphic rebuke to the philosophy of uniformitarianism, and gives remarkable support to the concepts of flood geology."[1]

The comparison of the deposits of rock debris and volcanic ash from the explosive eruption of Mt. St. Helens with the rock layers exposed in the Grand Canyon of Arizona is elaborated in greater detail in a newsletter article,[2] a videotape/DVD,[3] a companion print publication,[4] and a journal article.[5] Claims are made in some of those publications that Spirit Lake provides a likely environment for coal formation from the trees that were swept into the lake by the Mt. St. Helens eruption and aftermath.

So, taking all of the events that are/were associated with that eruption into account, what have we learned from the 1980 explosive eruption of Mt. St. Helens and its aftermath?

Uniqueness

The explosive eruption of Mt. St. Helens was but one of many known historical explosive volcanic eruptions. However, the buildup leading to the eruption, the eruption itself, and the aftereffects have been studied and observed in much greater detail than for any other such eruption, with the help of modern observing methods and tools. We note:

1. A major part of the explosive force was directed horizontally northward from the volcano. Other known explosive volcanic eruption forces were directed primarily upward, and such a large horizontally directed explosive force had never been observed before.

2. The eruption was accompanied by a major rock slide from the north side of the mountain, mixing rock slide debris with volcanic ash and pumice.

Non-uniqueness

Other than the two items noted above, the characteristics of the eruption were pretty much like those of other volcanic eruptions, explosive and non-explosive.

1. Volcanic ash was blasted high into the atmosphere from Mt. St. Helens, carried downwind, and deposited on the land surface as it cooled and settled out of the atmosphere. Such ash deposits are common, and are easily recognizable as products of volcanic activity. For example, just west of Kennewick, WA, two volcanic ash layers are observable within non-volcanic silt deposits, one traceable to

Glacier Peak, northwest of Kennewick, and the other traceable to Mt. Mazama (Crater Lake, OR), resulting from explosive eruptions of those Cascade Range volcanoes.

2. Mudflows are often observed in connection with volcanic activity. The resulting deposits are recognizable and easily identified. For example, mudflow deposits are found along the White River at Mt. Rainier, and at least one mudflow reached as far away as Enumclaw, northwest of the mountain.

3. Lahars (like mudflows, but with larger chunks of rock mixed with the mud and volcanic ash) are common in connection with volcanic eruptions, and are readily recognized as having been triggered by volcanic activity.

4. Volcanic deposits are often stratified, as, for example on the south slopes of Mt. Kilauea on Hawaii. However, the composition of the deposit readily identifies the source of the deposit as volcanic.

5. The erosion of terrain by mudflows and lahars is also commonly observed in connection with those events. Rapid erosion, to be sure, but otherwise unremarkable.

6. The erosion of the apron of rock slide debris and volcanic ash and pumice north of Mt. St. Helens since the eruption is very much like the erosion of any unconsolidated deposit by surface water. Take a look at the videocamera view on the Mt. St.Helens website (click on Google, type in "Mt. St. Helens," and select the "Volcanocams" item from the menu) when the weather is clear to see how the erosion channels have developed, many with high vertical channel walls.

So, what features of the erosion of the Mt. St. Helens area by mudflows, lahars, and continuing runoff of surface waters are comparable to the Grand Canyon of Arizona?

1. There is stratification (lamination) of the deposits through which the channels have been eroded in both the Grand Canyon and in the Mt. St. Helens area.

2. The channels were, and continue to be, eroded by natural processes much like those we observe in operation elsewhere today.

That's about it.

There are many differences between the Grand Canyon of Arizona and the channels that were formed in the Mt. St. Helens area during and following the 1980 eruptions.

1. The erosion at Mt. St. Helens was through loose, unconsolidated sediments; that of the Grand Canyon through solidified rock. The observation that the rock layers of the Grand Canyon can be matched from the north side to the south side of the Colorado River provides evidence that the rocks were solidified before the canyon was formed.

2. There is no evidence that mudflows or erosive lahars have contributed to the carving of the Grand Canyon by the Colorado River; there are no remnants of such events along the sides of the Canyon. There are rock falls from the cliffs of the Grand Canyon from time to time; these serve to widen the canyon. There have been debris flows that entered the Colorado River from some of the side canyons since the exploration by John Wesley Powell in 1869; these may have aided erosion of the side canyons, but they only impeded the flow of the Colorado River, delaying rather than aiding the downward erosion of the Inner Gorge. The major agent for erosion of the Colorado River channel appears to have been the erosive forces of sediments carried by running water, continuing today.

3. There are some lava flows that entered the Grand Canyon from volcanoes on the plateau north of the canyon; these flows occurred after the canyon had reached a stage close to its present form. Those lava flows blocked the erosion of the canyon for a time, and eventually the Colorado River eroded its channel through the lava flows. Remnants of the lava flows remain along the walls of the canyon. The carving of the canyon was done by the river, not by the lava flows.

4. There is much more to the Grand Canyon than the facts that the rocks are in layers, and that a canyon has been cut through them. The horizontal rock layers in the upper part of the Grand Canyon consist of some distinct layers of sandstone, interspersed with layers of shale, and both interspersed with layers of limestone. Below that,

in the tilted layers of the Grand Canyon Series, we find a layer of quartzite (an altered sandstone), layers of shale, layers of limestone, and some bedded lava flows. Below that, in the Inner Gorge, we find metamorphic rocks which contain evidence that they once were layers of sediments. Intruded through the metamorphic rocks and the Grand Canyon Series of tilted layers we find dikes of igneous granite.

There is none of that in the laminated deposits resulting from the Mt. St. Helens 1980 eruption and aftermath, but only rock slide rubble consisting of volcanic rock debris, and volcanic ash.

Spirit Lake

Spirit Lake, north of Mt. St. Helens, was profoundly affected by the 1980's eruption and aftermath. The studies of lake characteristics that were performed before 1980 indicated that Spirit Lake was very similar to other sub-alpine lakes in the region. The lake was very carefully monitored in detail after the 1980 eruptions.[6] Superheated volcanic rock, mudflows, ashfall, geothermal waters, and huge quantities of avalanche debris entered the lake within minutes of the eruption. The lake water was sloshed high on the slopes north of the lake boundary, and washed tens of thousands of trees and forest vegetation back into the lake with the receding water.

Following the eruption of Mt. St. Helens, a team of scientists arrived at the lake in summer 1980 to collect samples and take measurements of the lake characteristics. They found murky water, at a temperature 20°C above normal levels, with greatly elevated levels of dissolved minerals from the sediments and volcanic fluids that had entered the lake. They found almost no phytoplankton, the microscopic floating organisms with chlorophyll that were carrying on photosynthesis in the pre-eruption lake waters, providing the nutrients for virtually all other living organisms in the lake. The level of bacteria in the water was very high, including some toxic species. The water was almost completely devoid of dissolved oxygen, except near the surface, with bacterial action consuming nearly all of the oxygen in reactions with organic plant matter in the decay process. The recovery of the lake to its pre-eruption state seemed a far-distant possibility.

However, by the summer of 1981, after winter storms, rainfall, and snowmelt had brought fresh water into Spirit Lake, the levels of dissolved minerals and bacteria concentrations had diminished significantly, and oxygen content of the water had risen somewhat. By 1984, when the level of the lake surface had been stabilized by pumping, then opening a tunnel for drainage, the levels of dissolved minerals had decreased still more, and the lake had returned to its previous fully oxygenated state, except for partial depletion near the bottom sediments in late summer. By 1989 the clarity of the lake water had returned to nearly its pre-eruption value, and the most abundant organisms in the lake water were again the phytoplankton, though dominated by different species than in the pre-eruption lake.

With the lake water again fully oxygenated, the vegetation matter still in the lake may be expected to continue to decay, though the process will probably be slow in the cold lake water. It seems highly unlikely that the tree bark at the bottom of the lake will ever be transformed into coal, as was suggested as a possibility in the publications that promote the view that Earth is a recent creation.

Conclusion

If truth be known, the 1980 eruptions of Mt. St. Helens and the aftermath are a model for explosive volcanic eruptions and aftermath, nothing less and nothing more. A direct comparison between the rock slide deposits from the 1980 Mt. St. Helens volcanic eruption and the rock layers of the Grand Canyon is simply not appropriate because of major differences in rock type, and differences in resistance to erosion by those different rock types. A lot of pertinent information about those differences has been ignored and omitted from the publications that promote the view that Earth is a recent creation.

References

[1]Henry M. Morris, and John D. Morris, *Science, Scripture, and the Young Earth*, (El Cajon, CA: Institute for Creation Research, 1989), 30.

[2]Steven A. Austin, "Mount St. Helens and Catastrophism," *Acts & Facts* (El Cajon, CA: Institute for Creation Research, July 1986).

[3]Austin, *Mt. St. Helens: Explosive Evidence for Catastrophe* (Videotape/DVD) (El Cajon, CA: Institute for Creation Research, 1990).

[4]John D. Morris, and Steven A. Austin, *Footprints in the Ash.* (El Cajon, CA: Master Books, 2003).

[5]Harold G. Coffin, "Mount St. Helens and Spirit Lake," *Origins* 10, No. 1 (1983), 9-17.

[6]Douglas Larson, "The Recovery of Spirit Lake," *American Scientist* 81, (March-April 1993), 166-77.

7 Alaska Dinosaurs and Permineralized Bones

Several years ago some fossil bones of dinosaurs were discovered along the Colville River on the North Slope of Alaska. Most of the bones are of plant-eating hadrosaurs, although at least twelve different types have been found, including meat-eating predators. The fossils are found in the Prince Creek Formation, Cretaceous in age. Descriptions of the finds have been published in the scientific literature[1] and in Alaska State publications.[2]

The fossils are of special interest for a couple of reasons: 1) they are found farther north than any other known Cretaceous dinosaur fossils, and 2) some of the bones have undergone very little permineralization. The far north location gives rise to various questions about climate and food supply, but we won't deal with those questions here. We will pay attention to the matter of permineralization.

Most fossil bones of any appreciable age are thoroughly permineralized, that is, the open structure of the original bone material has been completely filled with rock-forming minerals, and often the bone material itself has been replaced by rock-forming minerals. The fact that some of the Alaska dinosaur bones are "very little" permineralized gives rise to interesting questions.

Most of the factors in the process of permineralization of fossil bones are pretty well understood. The usual process is thought to involve burial of the bones in porous sediments, and the soaking of those bones in groundwater carrying various rock-forming minerals in solution. The dissolved minerals are deposited in the bone cavities, and often the original bone material undergoes chemical reactions with the mineral-laden water and is carried away, at the same time that it is being replaced by other minerals. Sometimes the mineral replacement occurs in a way that preserves the bone structure in detail, with cell walls and other structural details of the bone clearly observable. The usual process is thought to be slow, requiring long periods of time for thorough permineralization to occur.

The news of the Alaska dinosaurs gained international attention. The bones were described in one publication as being

"as fresh as old cow bones."[3]

That description is a bit of an exaggeration, however; the smooth surfaces of the bones

"are stained a dark red brown."[4]

Roland Gangloff, Earth Science Curator at the Museum of the North of the University of Alaska at Fairbanks has collected and studied thousands of bones from the Colville site, and says that

"All of the bones are moderately to heavily permineralized. It varies with the bone bed and what part of the bed one samples. All of the bones have been highly saturated with iron oxides, and this is quite apparent to anyone by just looking at their dark brown color."[5]

Nevertheless, the low extent of permineralization in some of the bones is surprising.

When the news spread throughout the popular literature, advocates of the view that Earth is a recent creation seized the opportunity to claim that the less-permineralized state of these bones supports the idea that the Earth is young, as stated in the summary of an article on the subject:

"In summary, therefore:
"1. Most fossil dinosaur bones still contain the original bone.
"2. Even when heavily permineralized ('fossilized'), this does not need to require more than a few weeks."[6]

We should remember, however, that only some of the Alaska dinosaur bones are only partially permineralized, and that many are thoroughly permineralized. Does that mean that some of the bones are old, and some are young? That wouldn't make sense, since they were all found in the same rock formation.

If all the Alaska dinosaur bones are old, as the Cretaceous age of the surrounding rock formation would indicate, then why are some of them only partially permineralized? That is the interesting question. We know of another case in which some dinosaur bones are not permineralized; some of the fossil dinosaur bones found in the badlands of the Red Deer River Valley in Alberta, Canada are surrounded and encased by iron oxide, and the bone inside is not permineralized. The structure of the encasing iron oxide is so tight that groundwater is unable to penetrate to the inside for chemical reactions with the bone.

When confronted with the information that most fossil dinosaur bones, and most fossil bones of any type except the very recent ones, are permineralized, the advocates of the view that Earth is a recent creation respond that, well:

> "'Modern bones that fall into mineral springs can become permineralized within a matter of weeks.' So even a rock-solid, hard shiny fossil dinosaur bone, showing under the microscope that all available spaces have been totally filled with rock minerals, does not indicate that it *necessarily* took millions of years to form at all."[7]

I do not doubt that bones can become permineralized rapidly in mineral hot springs. The chemical environment of a mineral hot spring is quite unique. The question is: what does that mean with regard to the permineralization of fossil bones of dinosaurs and other bony fossils? Are all permineralized fossil bones found in mineral hot springs deposits? Most mineral hot springs build up deposits that are easily recognized as the products of hot spring activity. But that is not the kind of rock in which we find most fossil bones of dinosaurs; those bones are found mostly in sandstone and shale. There is no evidence that bones buried in silt and clay and sand become permineralized rapidly.

If truth be known, the degree of permineralization of fossil bones has never been associated with the passage of time in any way but broad generalizations. The process usually takes a long time. It will be interesting to find out what conditions might result in slowing or preventing the permineralization of bones as old as those Cretaceous dinosaur bones in Alaska. The fact that they are not completely permineralized, however, does not necessarily mean that they are recent. If truth be known, the fact that only some - not all - of them are only partially permineralized supports the conclusion that they are an exception, and the fact that most of the bones are permineralized argues against, rather than for, the idea that they are recent.

References

[1]Elisabeth M. Brouwers, William A. Clemens, Robert A. Spicer, Thomas A. Ager, L. David Carter, and William V. Sliter, "Dinosaurs on the North Slope, Alaska: High Latitude, Latest Cretaceous Environments," *Science* 237 (25 September 1987), 1608-10.

[2]L.J. Campbell, "The Terrible Lizards," *Alaska Geographic* 21 (1994), 24-37 (Bibliography, 108-9).

[3]Philip J. Currie and Eva B. Koppelhus, *101 Questions about Dinosaurs.* (Mineola, NY: Dover Publications, 1996), 12.

[4]Kyle L. Davies, "Duck-bill Dinosaurs (Hadrosauridae, Ornithischia) from the North Slope of Alaska," *Journal of Paleontology* 61, No.1 (1987), 198-200.

[5]Roland A. Gangloff, Personal communication (2002).

[6]Carl Wieland, "Dinosaur Bones: Just How Old are They Really?" *Creation ex Nihilo* 21, No. 1 (1999), 55.

[7]Wieland, "Bones," 54.

8 The Green River Formation: Lake Varves and Turbidity Flows

Varves

The sediments that are deposited on the bottoms of alpine and sub-alpine lakes that freeze over during winter display a distinct layered structure. The composition, texture, and general color of the sediments vary from season to season. The accumulation of sediments over a span of years consists of darker-colored and lighter-colored layers alternating in a regular cycle, with each pair of layers representing one year. These pairs of recurring layers are called "varves."[1]

Having observed this pattern, it makes good sense to measure the period of time during which such sedimentation has been going on in lakes that freeze over each winter by counting the varves in the bottom sediments. Many such measurements have been done, especially in lakes that were formed along with the melting of the glaciers that covered many regions of the Northern Hemisphere several thousand years ago.[2]

In many non-glacial lakes, a similar pattern of alternating layers results from a seasonal influx of clay and sand during spring and summer, alternating with phytoplankton growth that produces a darker colored, more carbon rich deposit of sediment during winter.[3]

One should also be able to get a pretty good idea of the passage of time in the sedimentation of lakes that do not freeze over every winter, if evidence indicates that they were annual deposits. One feature of annual deposits on the lake bottom would be uniform thickness of layers over long distances, indicating that the processes depositing the sediments were active over the entire area of the lake, and were not produced by local, shoreline events.

The Green River Formation

In southwestern Wyoming, northwestern Colorado, and eastern Utah there is a thick series of sediments in thin layers (laminae), collectively known as the Green River Formation. The Green River Formation sediments were deposited over a broad region that included a number of lakes surrounded by mountainous regions. There were three large lakes, given the names of "Fossil Lake," "Lake Gosiute," and "Lake Uinta," and many smaller lakes. Probably the best known sediments, frequently mentioned in popular literature, are those that were deposited in Fossil Lake in western Wyoming; Fossil Butte National Monument is located on these deposits. The lake bottom sediments include a large number of fossil fishes and other creatures that inhabit fresh water lakes, including mosquitoes and crocodiles. The shoreline deposits include cattails, fossil wading birds, and other creatures that live along lake shores in moist temperate or subtropical climates.[4]

At its greatest extent, Fossil Lake covered an area about 40-50 miles long from north to south, and 20 miles wide from east to west, according to the National Park Service brochure. The lake bottom deposit is made up of alternating layers, one lighter in color and consisting of inorganic sediment, followed by a thinner and darker layer rich in organic material. The conclusion has been drawn that each pair of alternating layers represents deposits laid down in one year. At its thickest, the lake bottom deposits of the Green River shale consist of between 6 and 7 million pairs of alternating layers. Those observations indicate that the duration of sedimentation in the lake was about 6 or 7 million years.[5,6] There is evidence of some cycles of deposition of longer duration in the Green River Formation, superimposed on the annual cycles, also accumulating to a total of millions of years.[7]

Old or young?

Various publications that promote the view that Earth is a recent creation have made claims disputing the reliability of counting varves or varve-like lake bottom deposits as a way of determining the ages of those deposits. The varve-like layers of the Green River Formation are often the target of such claims because of the long span of existence of the three largest lakes in the Green River Formation that is implied by the varve-like layers. For example, in *Science, Scripture and the Young Earth*, commenting on fossils found in the Green River Formation:

> "The fact that 'abundant' fossil fish and 'enormous concentrations' of fossil birds are found in the Green River formation surely ought to satisfy anyone that this is not a varved lake-bed at all,[2] but a site of intense catastrophism and rapid burial."[8]

And the footnote referred to by the superscript "2" in that quotation reads as follows:

> "Even in modern lakes, the so-called 'varves' may well be formed by catastrophic turbid water underflows, with many being formed oftener than annually. See A. Lambert and K. J. Hsu, 'Non-Annual Cycles of Varve-like Sedimentation in Walensee, Switzerland,' *Sedimentology* **26** (1979), 453-61."[9]

So let us examine those claims in more detail, along with the papers from the professional scientific literature that are referred to in connection with the claims. We'll deal with varves and varve-like deposits in this chapter, and come back to the fossils in the next chapter.

Lake varves and turbidity flows

The paper by Lambert and Hsü, referred to in *Science, Scripture, and the Young Earth* as noted above, reports a study of sediment deposits in Lake Walensee, Switzerland.[10] The authors observed that there are occasional turbidity flows that are deposited on the lake bottom, and they describe one such turbidity flow in some detail.

A turbidity flow consists of a slurry of sediments and water, possibly generated by a slumping of unconsolidated sediments on the steeper slopes along the edge of the body of water. This slurry is more dense than clear water, so it flows along the bottom. The slurry may continue to flow for some distance after reaching the flatter bottom of the body of water farther from the shore, carried by the momentum generated by the force of gravity on the steeper slope. Turbidity flows are observed to occur from time to time along the continental slope in the world's oceans, as well as in lakes.

Of course, the sediment carried by such a turbidity flow adds a layer to the bottom of the lake which is not due to the annual variation in the varve or varve-like sequence. The claim in *Science, Scripture, and the Young Earth* is that such turbidity flows would introduce error into any attempt to measure the passage of time by counting varve-like layers in lake deposits, particularly in the Green River Formation. So we should take a brief look at that matter.

The turbidity flow in Lake Walensee described by Lambert and Hsü was observed to progress over the floor of the lake at a speed of 50 cm. per second. Converted to more common speed units, that's a bit more than one (1) mile per hour, the speed of a slow sauntering walk. Hardly catastrophic, but it does add a layer to the bottom sediment that is not due to annual varve-like sedimentation. In other lakes studied by Hsü and Lambert, some additional layers of sediment were deposited by turbidity flows associated with flooding of streams that empty into the lake following heavy rains.

So, yes, layers of sediment may be added to lake bottom deposits by turbidity flows from slumps along the margins of the lake, or from unusually heavy input of sediment from flooding streams that enter the lake. However, turbidity flows usually deposit layers of sediment with larger particles on the bottom of the layer and progressively smaller particles toward the top. This is called "graded bedding," and it occurs because the larger particles settle out first, and are followed by progressively smaller particles. Most turbidite beds can be recognized by this graded bedding.

It is usually possible to distinguish between turbidity flow deposits and true varves or varve-like annual deposits, so that a counting of the varves or varve-like annual layers would still be reliable. The turbidity flow deposits would be recognized as such and would not be included in the count. Of course, it requires an understanding of the processes involved, and appropriate attention to details, to get reliable results; that's true for any investigation of anything.

The Green River Shale

The laminations in the Green River Formation are not true varves; the lakes in which the sediments were deposited were not alpine or sub-alpine lakes, and they did not freeze over during winter.

Are the laminations in the Green River Formation annual layers? There is a considerable amount of evidence to support the conclusion that they are annual layers. The individual laminae are consistent in thickness and can be correlated in core samples over long distances, sometimes distances of several tens of kilometers.[11] Furthermore, the graded bedding that is characteristic of turbidity flows is NOT found in the thin-bedded laminae of the lake bottom deposits in the Green River Formation. Because turbidity flows slow down as they progress, and the sediment deposits of turbidity flows are progressively thinner and finer-grained toward the end of the flow, it is unlikely that turbidity flows would deposit consistently thin-bedded layers over the large distances and broad areas occupied by the Green River shales.

Conclusion

If truth be known, the record of sediments in the Green River shale testify to the millions of years of history of the lakes and sedimentation processes occurring in that part of the world. That history is only a small portion of the total history of the Earth, of course.

Again, you may choose to believe that the Earth is a recent creation for other reasons, if you wish, but the record of the varve-like lake bottom deposits in the Green River Formation does not provide support for that perspective.

References

[1]Kenneth J. Hsu, *Physical Principles of Sedimentology*, (New York: Springer-Verlag, 1990), 41.

[2]Gunilla Petterson, "Varved sediments in Sweden: a brief review," in *Palaeoclimatology and Palaeoceanography from Laminated Sediments*, A.E.S. Kemp, Ed., Geological Society Special Publication No. 116. (London: The Geological Society, 1996), 73-77.

[3]Maurice E. Tucker, *Sedimentary Petrology: An Introduction to the Origin of Sedimentary Rocks*. (Oxford: Blackwell Science, 2001), 105.

[4]W.H. Bradley, "Geology of the Green River Formation and associated Eocene rocks in southwestern Wyoming and adjacent parts of Colorado and Utah," *U.S. Geological Survey Professional Paper 496-B* (Washington: Gov't Printing Office, 1964).

[5]Bradley, "The Varves and Climate of the Green River Epoch," *U.S. Geological Survey Professional Paper 158E* (Washington: Gov't Printing Office, 1929), 107.

[6]Bradley, "Origin and Microfossils of the Oil Shale of the Green River Formation of Colorado and Utah," *U.S. Geological Survey Professional Paper 168* (Washington: Gov't Printing Office, 1931), 26.

[7]A.G. Fischer and L.T. Roberts, "Cyclicity in the Green River Formation (lacustrine Eocene) of Wyoming," *Journal of Sedimentary Research* 61 (1991), 1146-54.

[8]Henry M. Morris and John D. Morris, *Science, Scripture, and the Young Earth* (El Cajon, CA: Institute for Creation Research, 1989), 34.

[9] Morris, *Science, Scripture*, 34.

[10]A. Lambert and K.J. Hsu, "Non-Annual Cycles of Varve-like Sedimentation in Walensee, Switzerland," *Sedimentology* 26 (1979),

453-61.

[11]John R. Dyni, "Regarding the laminations ("varves") in Green River Oil Shale representing annual events," (on website www.indiana.edu/~ensiweb/lessons/varve.ev.pdf, 2000).

9 Quotations and Misquotations: The Green River Formation

Let us return to the matter of the fossils that are found in the sedimentary rocks of the Green River Formation. We will evaluate the claim that:

"The fact that 'abundant' fossil fish and 'enormous concentrations' of fossil birds are found in the Green River formation surely ought to satisfy anyone that this is not a varved lake-bed at all, but a site of intense catastrophism and rapid burial."[1]

We will pay particular attention to the quotations from the professional scientific literature that are submitted as support for that claim.

Expectations in quotations from other publications

It is common practice to use quotations from other authors in articles in journals, in discussions in books, in public lectures, etc. It is entirely proper to buttress published ideas with quotations from other authors who are recognized as authorities in their field of expertise, or to use quotations from other authors in a publication presenting arguments against the claims and ideas of those other authors.

However, there are certain expectations of the reader, and of the author and publication from which the quotation is taken, that ought to be observed in the presentation and use of such quotations. Those expectations, or commonly accepted standards, include the following:

1. The quotation should be enclosed in quotation marks ("...") or set off from adjacent text in ways that clearly identify it as a quotation from another work.

2. The quotation should be a *verbatim* copy of the original publication. If some parts of the quoted section of the original publication are being omitted for the sake of brevity, that should be indicated by entering a series of dots (...) or equivalent notation, and the meaning of the original passage must not be changed by such an omission.

3. A paraphrase should be identified as such, and the meaning of the paraphrase should be consistent with the meaning of the original publication.

4. Any quotation will necessarily be lifted out of its context, but the quotation should not be used in an attempt to buttress ideas which are inconsistent with the context of the original publication.

Publications promoting the view that Earth is a recent creation often include quotations from recognized and respected scientists to buttress their claims of scientific support for the view being promoted. Unfortunately, the use of quotations in those publications does not always comply with the commonly accepted standards for using such quotations in published literature.

Fossils in the Green River Formation

The claim from *Science, Scripture, and the Young Earth* at the beginning of this chapter follows two quotations from the scientific literature, one about fossil fish in the Green River formation by H. Paul Buchheim and Ronald C. Surdam, and one about fossil birds in the Green River Formation by Alan Feduccia. We will consider each of these publications to see what they actually say.

The first thing we should note about the "abundant" fossil fish and the "enormous concentrations" of fossil birds is that the fish fossils are found in lake bottom sediments, and the bird fossils are found in shoreline sediments; the fish fossils and bird fossils are NOT found together. If both birds and fish had been deposited and fossilized in an event of "intense catastrophism," we would expect them to have been deposited together, I would think. But they are not. The birds died in an environment normally occupied by shore birds, and the fish died in an environment normally occupied by fish.

Fossil fish

The paper by Buchheim and Surdam is entitled "Fossil Catfish and the Depositional Environment of the Green River Formation, Wyoming." A passage from that paper is quoted (in italics) in *Science, Scripture, and the Young Earth* as follows:

"Furthermore, fossil catfish are distributed in the Green River basin over an area of 16,000 km² ....The catfish range in length from 11 to 24 cm., with a mean of 18 cm. Preservation is excellent. In some specimens, even the skin and other soft parts, including the adipose fin, are well preserved."[2]

In the research paper by Buchheim and Surdam, the description of the fossil fish and their occurrence continues beyond what was quoted in *Science, Scripture, and the Young Earth*. After an intervening paragraph, we read:

"The abundant and widespread occurrence of skeletons of bottom feeders, some with soft fleshy skin intact, strongly suggests that the catfish were a resident population. It is highly improbable that the catfish could have been transported to their site of fossilization. Experiments and observations made on various species of fish have shown that fish decompose and disarticulate after only very short distances of transport."[3]

That description doesn't sound like anything close to "intense catastrophism," and it explicitly rules out transportation to the site by the turbulent flow usually associated with massive flooding. Rather, the quotation from Buchheim and Surdam indicates that those fish died and were fossilized where they had lived.

The quotation in *Science, Scripture, and the Young Earth* was lifted out of its context. The quotation supposedly supports the claim of "intense catastrophism" at the site where the fossilized fish are found, but the quotation doesn't tell the whole story. It has left out some very important information regarding the question at hand. The context - the more complete text in Buchheim and Surdam - supports a position which is exactly the opposite of the claim in Morris.

Fossil shore birds

The title of the paper by Feduccia is "*Presbyornis* and the Evolution of Ducks and Flamingoes." The following passage is quoted (in italics) in *Science, Scripture, and the Young Earth*:

> "*Because bird bones are hollow or pneumatic as an adaptation for flight, they are not well preserved in the fossil record....During the early to mid-1970's enormous concentrations of Presbyornis have been discovered in the Green River Formation.*"[4]

The second part of the quotation above, the portion following the "....", is taken from a sub-section of Feduccia's paper entitled "The paleoenvironment of *Presbyornis*." So let us read a little further in that section, where we find the following:

> "*Presbyornis* is now known from eight or more localities, including tracks actually showing the webbed feet from a site in Utah. Perhaps the most spectacular locality is near the Colorado-Wyoming border, approximately 110 km. south of Rock Springs, Wyoming. Here the bones are found just below the rim of an enormous butte (Figs. 1 and 2), where, literally by the thousands, they protrude from the coarse sandstone that apparently eroded off the Uinta Mountains to the south. Henry Roehler has done considerable work on the geology of this region, known, rather cumbersomely, as the Canyon Creek facies of the Wilkins Peak Member of the Green River Formation. The sediments probably represent an extensive beach deposit that extended into the large Eocene Lake Gosiute.[5]

The caption to Fig. 2 in Feduccia's paper reads as follows:

'A block of matrix from the Canyon Creek locality illustrates the dense concentrations of *Presbyornis* bones. Even with such concentrations of fossils there is no reason to assume that a catastrophe suddenly destroyed great numbers of the birds; normal attrition within a large population of the highly colonial *Presbyornis* could easily account for such a dense collection of bones in the bottom of a lake.'

And, a bit further on in Feduccia's paper:

"Approximately 160 km to the north, large numbers of *Presbyornis* fossils occur at an area where the Wilkins Peak Member of the Green River Formation intertongues with the Cathedral Bluffs Tongue of the Wasatch Formation. This site, originally discovered by Faroy Simnacher and later excavated by McGrew, represents a different type of facies [that is, rock type], composed of a grayish-green silty claystone approximately 30 meters below the basal oil shales of the Laney Shale Member of the Green River Formation. The many egg shells in the deposit indicate that it was almost certainly a large nesting colony near the shoreline of a lake. The highly colonial *Presbyornis* thus flourished in habitats similar to those preferred by modern flamingoes."[6]

Note that the first site of bird fossils mentioned above is a "coarse sandstone," and the second site is "silty claystone," both of which indicate a shore deposit. The bird fossils are not found in the thin laminae of the varve-like deposits of lake bottom sediments.

In this case again, the quotation in *Science, Scripture, and the Young Earth* has been lifted out of its context. It was inserted into *Science, Scripture, and the Young Earth* as supposed support for the view that the birds were deposited in an event of "intense catastrophism," but the context - the more complete text in the paper by Feduccia, and especially the presence with the bones of tracks of webbed feet in one location, and egg shells in another - supports a position that is exactly the opposite of that promoted in *Science, Scripture, and the Young Earth*.

CLARENCE MENNINGA

Conclusion

So, the paper by Feduccia and the paper by Buchheim and Surdam present evidence and conclusions that both the fish and the birds were inhabitants of the environment of the lake, the fish in the water and the birds along the shore, and their remains were deposited and fossilized at the site where they had lived. This conclusion is exactly the opposite of the claim in the publications that promote the view that Earth is a recent creation, namely, that these fossils were "washed" into the area from somewhere else by catastrophic flooding.

As you see, the papers and authors of the scientific articles have been grossly misrepresented in the publications that promote the view that Earth is a recent creation, with the quotations from those papers taken out of their original context, and key parts of the articles omitted, in submitting them as supposed support for the perspective that Earth is a recent creation.

You may choose to believe that the Earth is a recent creation for other reasons, if you wish. But the evidence from the fossil fish and birds in the Green River Formation does not support that belief.

Mark Twain comments on geologists

Mark Twain (Samuel L. Clemens) was a humorist. His lectures during his lifetime and his writings have entertained millions of listeners and readers. One of those stories can provide an analogy for the treatment of scientific data by some of the publications that promote the view that Earth is a recent creation.

In his youth, Mark Twain was a cub pilot on Mississippi River steamboats. His stories from that experience are told in *Life on the Mississippi*, first published in 1883. In a chapter in that book, he tells the reader about meanders and cutoffs on the Mississippi River.

From Cairo, Illinois southward to the Gulf of Mexico, the river cuts a channel that is far from straight, but curves in great meander loops extending miles from side to side. The river erodes the outside

80

of the curves, and deposits clay and sand and silt on the inside of the curves. Over a period of time, the neck of the meander may become very narrow, and, at high water, the stream may flow over that neck and rapidly erode a new section of channel, cutting the meander off from the flow of the river. Mark Twain describes in detail several of the cutoffs that occurred during his lifetime, and reports additional cutoffs from earlier historical records. Each of those cutoffs has shortened the channel of the river by a known number of miles.

Then Mark Twain uses the data he has collected from recorded cutoffs on the Mississippi to poke fun at geologists. He wrote:

"Now, if I wanted to be one of those ponderous scientific people, and 'let on' to prove what had occurred in the remote past by what had occurred in a given time in the recent past, or what will occur in the far future by what has occurred in late years, what an opportunity is here! Geology never had such a chance, nor such exact data to argue from! Nor 'development of species,' either! Glacial epochs are great things, but they are vague - vague. Please observe:

"In the space of one hundred and seventy-six years the Lower Mississippi has shortened itself two hundred and forty-two miles. That is an average of a trifle over one mile and a third per year. Therefore, any calm person, who is not blind or idiotic, can see that in the Old Oölitic Silurian Period, just a million years ago next November, the Lower Mississippi River was upward of one million three hundred thousand miles long, and stuck out over the Gulf of Mexico like a fishing rod. And by the same token any person can see that seven hundred and forty-two years from now the Lower Mississippi will be only a mile and three-quarters long, and Cairo and New Orleans will have joined their streets together, and be plodding comfortably along under a single mayor and a mutual board of aldermen. There is something fascinating about science. One gets such wholesale returns of conjecture out of such a trifling investment of fact."[7]

And we all have a good laugh.

But let's take a bit more detailed look at Mark Twain's conclusions from his set of data on cutoffs. Those conclusions are pretty ridiculous, don't you think? By Mark Twain's data and reasoning, even three thousand years ago, about the time of the reign of King David over ancient Israel, the Mississippi River, if extended directly southward, would have reached across the Gulf of Mexico, across Central America at Guatemala, and into the eastern Pacific Ocean beyond the equator. Surely that wasn't the way it was! And the straight line distance from Cairo, Illinois to New Orleans, Louisiana is not getting any shorter as time goes by; those two municipalities will never merge into one.

So what's going on? Was Mark Twain's reasoning faulty? Were his data mistaken?

The whole story

Remember, Mark Twain was a humorist. He wanted to get a laugh from his audience, and, in this passage from *Life on the Mississippi*, he surely succeeded. But he didn't tell us the whole story. He wasn't obligated to do so; he was a humorist, and he wanted to get a laugh. So where is the flaw that produced such ridiculous conclusions? Well, he didn't include all of the available data as the basis for his reasoning. Remember? At the same time that the Mississippi River is cutting off a meander from time to time, it is always eroding its banks on the outside of the curves of the meander loops. The cutoffs shorten the channel, perhaps by several miles, in a very short time. The erosion of the outside of the curves of the meanders lengthens the channel, little by little, but continuously. On average, the length of the channel remains about the same.

Very clever! Very funny! Also very misleading, if you take Mark Twain's data and reasoning as if he had included <u>all</u> of the data that are important to the question of changes in the length of the Mississippi River channel.

There is far too much in the publications that promote the view that Earth is a recent creation that follows the pattern of Mark Twain's story of cutoffs; the reader is not told the whole story. The examples of fish and bird fossils in the Green River shale, related above, are just two of many. But the publications that promote the view that Earth is a recent creation are not trying to be humorous.

Differences of perspective, and consideration of alternate explanations for what we observe in God's world, are part of the fabric of science (and of life). But there is no rightful place in science (or in life) for misrepresentation.

References

[1] Henry M. Morris and John D. Morris, *Science, Scripture and the Young Earth.* (El Cajon, CA: Institute for Creation Research, 1989), 34.

[2] Morris, *Science, Scripture*, 33.

[3] H. Paul Buchheim and Ronald C. Surdam, "Fossil catfish and the depositional environment of the Green River Formation, Wyoming," *Geology* 5 (1977), 196.

[4] Morris, *Science, Scripture*, 34.

[5] Alan Feduccia, "Presbyornis and the Evolution of Ducks and Flamingoes," *American Scientist* 66, No. 3 (May-June 1978), 298.

[6] Feduccia, "Presbyornis," 301.

[7] Mark Twain (Samuel L. Clemens), *Life on the Mississippi,* (New York: P.F. Collier & Son Company, 1883), 155-6.

10 Petrified Forests of Yellowstone National Park

For many years the standard textbooks on historical geology featured the example of petrified trees and logs at Specimen Ridge and Amethyst Mountain in Yellowstone National Park as evidence for a long history of the Earth, and evidence for the development of our present landscapes by a sequence of natural processes. Early reports,[1] oft repeated in textbooks,[2] claimed a succession of 27 forests, one on top of the other, each buried by deposits of volcanic ash and breccias (broken pieces of volcanic rock in finer sedimentary matrix), and standing trees turned to stone where they had grown. The claim was that such a phenomenon could only mean that each forest had taken root and grown in soil on top of the previously buried forest, thus demonstrating the long time period of natural processes that produced what we see today.

More recently, new studies of the petrified stumps and logs and other fossilized plant material in Yellowstone National Park has brought that textbook scenario into question, with proposed explanations that some (or all?) of the fossilized plant material had been transported to its present site by mudflows and lahars.[3,4] (A lahar is like a mudflow, but with larger pieces of rock carried along with the mud.)

The re-study of the fossil forests in Yellowstone National Park has appeared in publications that promote the view that Earth is a recent creation, along with the claim that the entire sequence of strata in which the petrified trees occur could have been deposited in a very short time.[5,6]

Comparison with Mt. St. Helens

The 1980 eruptions of Mt. St. Helens, with the accompanying

mudflows, lahars, and results in the surrounding forested area, provided opportunity to make comparisons with the petrified forests of Yellowstone National Park. Many similarities were observed.

Thousands of trees and stumps had been washed into Spirit Lake, north of Mt. St. Helens, by the 1980 explosive eruption and aftermath. Some of the water-logged trees and stumps sank to the bottom of the lake, some of them in upright position. The mudflows had carried logs and tree stumps along, and some of the logs and stumps were deposited in the valley of the North Toutle River, and some in the standing forests along the way. The lahars had carried water, rock debris, volcanic ash, and plant material down the slopes of the mountain, and left the logs and stumps, along with the sediments, at the foot of the slope. All of this happened very quickly, in a matter of minutes or hours. Over the intervening years, more logs and stumps have settled to the bottom of Spirit Lake.

Interpretation

So, was the formation of the fossilized forests at Yellowstone the result of a sudden, violent, and short-lived event similar to the 1980 eruptions of Mt. St. Helens, only on a larger scale? There has been a considerable amount of recent research and discussion in the scientific literature regarding the formation of the Yellowstone fossil trees.

The publications that promote the view that Earth is a recent creation point out the possibility of transported trees and stumps being found with some in upright position, as on the bottom of Spirit Lake north of Mt. St. Helens. However, while some (estimated 10%) of the upright Yellowstone trees are found on sediments that were probably lake or pond deposits, most are found on terrestrial sediments.

Some of the logs and upright stumps found in the Yellowstone fossil forests are now recognized as having been transported from elsewhere. The paper by Fritz (ref. 4) was focused primarily on the transportation of leaves, needles, cones, pollen, and spores from somewhere at higher elevation, but included the recognition that some stumps and logs also gave evidence of having been transported from elsewhere. However, the paper included the statement that "some upright trees with long trunks and well-preserved root systems were apparently buried and preserved in situ." [that is, where they had grown][7] The scientific literature has pointed out the similarity of the Yellowstone fossil forests to the events at Mt. St. Helens in 1980; at Mt. St. Helens, also, some logs and stumps were carried by mudflows and lahars, and were deposited among the standing trees of the surrounding forest; those transported logs and stumps are now found along with and among standing, living trees.

Today there is virtually universal agreement in the community of professional scientists that many of the petrified trees at Yellowstone are found in deposits of sediments made up of fragments of volcanic rock cemented together in a finer-grained matrix, very likely as a result of lahars from surrounding volcanoes. The question is whether these lahars were all part of a single, sudden and short-lived event, or the result of many volcanic events over a significant period of time. Some observations are worthy of note:

1. There are numerous beds of volcanic ash interspersed throughout the sequence of lahar deposits at Yellowstone National Park. While volcanic ash is certainly produced in violent explosive eruptions of volcanoes, the deposition of that ash in bedded layers over the surrounding area is not the result of violent events. The ash quietly settles out from the atmosphere over a period of time onto a land surface, or perhaps into a quiet body of standing water. It seems unlikely that successive lahars could be separated by ash beds if those successive lahars were not separated by an appreciable extent of time. The deposition of the multiple layers of ash interbedded with lahar deposits most certainly cannot have taken place during a single, turbulent, worldwide flood.

2. There are also numerous preserved soil layers, or organic-rich layers, on which the standing petrified trees at Yellowstone are found

in growth position. These layers are found interspersed among the various layers of lahar deposits. Only a small fraction (about 10%) of these soil layers provides evidence of having been formed in lakes or ponds. It seems highly unlikely that any such soil or organic-rich layers could be formed if the numerous lahars had followed one another in rapid succession.

3. The various levels of fossilized trees lie one on top of the other, each with its fossilized leaves, needles, cones, pollen, and spores, each with some stumps and logs that were transported from elsewhere, and each with some upright trees preserved where they had grown. Each level represents an event somewhat similar to the results of the 1980 eruption at Mt. St. Helens. The total sequence of levels at Yellowstone represents many such events, each deposit being laid down on top of the previous ones, with volcanic ash falls, soil formation, new forest growth, etc., between layers. It hardly seems plausible to suppose that such a sequence of events would occur in a brief span of time.

The consensus in the scientific literature is that the petrified forests of Yellowstone National Park really ARE a succession of forests, though with a more complex history than had been accepted earlier.

Conclusion

Of course, you may choose to believe that the Earth is a recent creation, formed in all its complexity over a short period of time, but the petrified forests of Specimen Ridge and Amethyst Mountain in Yellowstone National Park do not provide scientific support for that belief.

References

[1]Erling Dorf, "The petrified forests of Yellowstone Park," *Scientific American* 210 (1964), 106-14.

[2]Carl O. Dunbar and Karl M. Waage, *Historical Geology*, 3rd Ed. (New

York: John Wiley & Sons, Inc., 1969), 421.

[3]Harold G. Coffin, "Orientation of trees in the Yellowstone petrified forests," *Journal of Paleontology* 50, (1976), 539-43.

[4]W.J. Fritz, "Reinterpretation of the depositional environment of the Yellowstone 'fossil forests,'" *Geology* 8, (1980), 309-13.

[5]Coffin, "The Yellowstone Petrified 'Forests,'" *Origins* 24, No. 1, (1997), 2-44.

[6]John D. Morris and Steven A. Austin, *Footprints in the Ash.* (El Cajon, CA: Master Books, 2003), 100-2.

[7] Fritz, "Reinterpretation," 313.

11 Falsehood in Science: Piltdown Man

Deception and Fraud

Nobody likes to be deceived. We especially dislike losing money to a con man or a scam. Yet, stories abound of those unwitting souls who are taken in by deceit and falsehood. Those of us who have email addresses are in a constant battle to avoid being taken in by the scams on the electronic highway. Fraud is rampant in our modern world, and has been part of the context of our lives for as long as history.

Some of those cases of fraud have been in the headlines of our newspapers. They may involve high-ranking managers of multibillion dollar corporations, or the secretary in a local government or church office. They may be in the form of cheating in schools, at all levels, including prestigious colleges and universities. Some cases involve preachers whose lives are inconsistent with the Christian gospel they preach from week to week. Reporters for major newspapers have invented stories and reported them as fact, and had them published; the editors of their papers were embarrassed when the falsehoods were exposed. Some people apply for a job and submit false claims of academic credit or experience on their resumé. Reports of those kinds of deceptions and frauds and more are much too common in our daily newspapers. When the motives for fraud and deceit can be identified, they include greed, pride, a false sense of "saving face," and a host of complex moral failings. If we are honest with ourselves, all of us must confess that we sometimes face opportunities to cheat and defraud others, and we pray and struggle to keep ourselves from falling into those temptations.

With fraud so common in everyday life around us, it should not come as a surprise to find out that there is fraud in science, as well. A number of examples were published in a relatively recent book,[1] and I will tell one of the stories in some detail. It was in all the newspapers and news magazines, and attained national attention.

Known as the "Summerlin Affair," or, more descriptively as "the painted mouse," the attempt at deceit took place at the prestigious Sloan Kettering Institute in New York in 1974. Doctor/scientist William T. Summerlin had been seeking ways to transplant tissue from one individual to another unrelated individual and avoid having the tissue rejected by the recipient's immune system. Although considerable progress has been made in improving acceptance of transplanted tissue since the Summerlin Affair, experience at that time indicated that tissue transplantation is generally unsuccessful unless the donor and recipient are genetically identical, as they are in a strain of purebred mice or in human identical twins.

Summerlin was attempting to reduce the likelihood of transplanted tissue rejection with experiments in mice. He obtained skin from black mice, held the skin in nutrient solutions outside the donor's body for a period of time, and then transplanted the skin onto white mice. The treatment of holding the skin in nutrient solution for a period of time was based on the supposition that such treatment would remove from the cell surface those biochemical antigens that were known to attract attack by the recipient's immune system and lead to tissue rejection. Summerlin had claimed success in using that procedure in previous experiments, and he transplanted some patches of skin from the black mice to the genetically unrelated white mice according to that procedure.

On the morning of 26 March 1974, Summerlin carried some of his mice with transplanted skin to the office of Dr. Robert Good, Director of the Institute, to demonstrate his success. While in the elevator on the way, he used his black felt-tip pen to darken an area of the transplanted skin on two white mice in order to make it look like a more successful transplant than it really was. Later that day, after the mice had been returned to the animal room, a laboratory assistant noticed that the grafts looked "unusual." He found that the black color could be washed away with alcohol, and he reported his findings to a senior technician, and thence up the chain of command to Director Good. A peer review committee was appointed to investigate the matter.

The committee produced a report on 20 May 1974. They confirmed that Summerlin had misrepresented his findings and observations, and they also noted that some his past claims of successful tissue transplants were suspicious. They expressed the belief that:

> "some actions of Dr. Summerlin over a considerable period of time were not those of a responsible scientist"

and that his

> "irresponsible conduct is incompatible with discharge of his responsibilities in the scientific community."[2]

The case was widely publicized. In his analysis of the case, Kohn (ref 1) spreads some blame around to others, without excusing Summerlin for his attempt at deceit.

There are some common strands of circumstances surrounding many of the known cases of attempted fraud in science. A lack of sufficiently close supervision by superiors is often suggested as contributing to the opportunity for deceit. There is a nearly universal pressure on research scientists in all fields to produce positive results. Positive results are often needed to renew or obtain grants to support the research. There are also opportunities to gain understanding from negative results of scientific experiments, but negative results cannot be patented or sold. There are also personal pressures on scientists at all levels within a research institution to publish papers in the professional literature. Prestige, recognition by peers, honors, and often salary increases and advancement are on the line.

From time to time someone in science succumbs to the pressure. Is that basically different from attempts at financial fraud?

We should remember that the broad enterprise known as "science" is not discredited by the cheaters, just as the banking industry is not discredited by embezzlers, or the clergy by the imposters, or the medical profession by the quacks. An editor of the journal *The Scientist* wrote:

"It would be naïve to expect that science should differ from other crafts in not including its aliquot of bent practitioners."[3]

The Famous Piltdown Hoax

One of the most famous forgeries in science is the case known as "Piltdown Man," or the "Piltdown Hoax." Several books have been written about it. One of those, by Charles Blinderman,[4] is the primary source for the following synopsis. The story has even found its way into the publications that promote the view that Earth is a recent creation, including a book[5] and a website.[6]

The fraud took place in southeastern England during the early years of the 20th Century. Charles Dawson, a central figure in the story, was a local lawyer in the region of Sussex. He was also a collector of antiquities (fossils), and an amateur geologist. He had received credit for the discovery of three new fossil species, was author of six books on local histories, and had been designated Honorary Collector for the British Museum of Natural History.

In 1908 a piece of human skull was found in a gravel pit on the property of Barkham Manor (farm) at Piltdown. The fragment was probably found by workmen, and was given to Charles Dawson. Three years later, in 1911, Dawson went collecting at the gravel pit, and found additional pieces of human skull in the refuse heaps at the pit. In February 1912 Dawson wrote a letter to Arthur Smith Woodward, Keeper of Geology at the British Museum of Natural History, telling him about the skull fragments. Later that year Dawson found a broken section of a right lower jawbone with two intact molar teeth in the same pit. The jaw had some characteristics similar to the jaw of chimpanzee, but positive identification was difficult, because the condyle, the part of the jawbone that would have formed the joint with its cranium, was broken off, the chin was missing, and the two molars appeared to be considerably worn. Also found in the pit in 1912 were some fossil fragments of hippopotamus, mastodon, red deer, horse, and beaver, and some iron-stained flint tool fragments. The human skull pieces, the jawbone, the other fossils, and the flints were all stained brown, as if they had lain in the iron-rich gravels for a long time.

Two questions arose immediately: 1) How old were the human cranium fragments? and 2) Did the jaw found in the pit belong to the same individual as the skull pieces?

On May 23, 1912 Dawson wrote a letter to Woodward at the British Museum of Natural History to announce that he was coming to London, and he brought the fossils with him soon thereafter. The pieces of the cranium were put together to reconstruct the original skull, and the jaw was accepted by Woodward (and others) as belonging to the same individual. By now, excavators at the site included Woodward and Father Pierre Teilhard du Chardin. The announcement of finding a human skull with ape-like jaw appeared in the Manchester Guardian on November 21, 1912, and the reconstructed specimen was displayed publicly in December of that year.

Doubts were expressed immediately, but the roster of those who defended the authenticity of the reconstruction, with jaw and cranium belonging to the same individual, constitutes an impressive list of well-known and highly respected authorities in their fields. Further support came from the Piltdown gravel pit in September 1913 with the find of a canine tooth, apparently from the jaw of a primate, midway in size between that of a chimpanzee and an adult human. A Royal Academy portrait was taken in 1915 portraying the men involved in the find and reconstruction of Piltdown Man, along with the reconstructed skull. Those pictured are: Arthur Smith Woodward, Keeper of Geology at the British Museum of Natural History, Fellow of the Royal Society, and Vice-President of the

Geological Society; F. O. Barlow, Woodward's assistant at the Museum; Grafton Eliot Smith, Professor of Anatomy, Victoria University, Manchester; Charles Dawson, the discoverer of the fossils; Arthur Keith, Conservator of the Royal College of Surgeons; W. P. Pycraft, Curator of Anthropology at the Museum; Arthur S. Underwood, Professor; and Edwin Ray Lankester, Director of the British Museum of Natural History. The announced consensus of these eminent British scientists was that the skull and jaw represented a single individual, and named the new species *Eoanthropus dawsoni*.

The age of the Piltdown specimen, based on the other fossils found in the gravel pit, was judged to be somewhere between early and middle Pleistocene, perhaps as much as a million years. Ancient hominid specimens had been found in other parts of Europe, and now it appeared that Britain had its own ancient hominid, at least as old as those that had been found on the Continent.

There were early dissenters on both sides of the Atlantic. David Waterston, Professor of Anatomy at King's College, compared radiographs of the Piltdown jaw with the jaw of a chimpanzee, and his conclusion that the jaw was simply that of an ape was published in the journal *Nature* in November 1913. A dental anatomist, Dr. Courtney W. Lyne, presented a paper at the meeting of the Odontological Section of the Royal Society on January 24, 1916 in which he argued that the canine tooth belonged to a youthful individual, while the cranium was clearly that of an older adult, and that the order in which the teeth had irrupted in the Piltdown jaw was like that of the chimpanzee, but unlike that of humans. Gerrit S. Miller, a research scientist at the Smithsonian Institution in the United States published a paper in *Smithsonian Miscellaneous Collections*, Vol. 65 (November 24, 1915) in which he presented detailed anatomical measurements of a cast of the Piltdown jaw, and concluded that the jaw was that of an ape. The defenders published their arguments on the pages of the British journal *Nature*, while the American dissenters published their objections on the pages of the U.S. journal *Science*. American naturalist Roy Chapman Andrews, writing during the span of time covered by the discussion, wrote:

"The problem of Piltdown Man has addled the best scientific brains for thirty-five years. Few authorities have agreed about much of anything concerning it, particularly not its evolutionary position or relationship. Dr. Franz Weidenreich, in his 1943 monograph,[7] throws it out of the line of very primitive types entirely. He believes it is *Homo sapiens*, modern man, and that the jaw cannot possibly belong to the skull no matter what were the circumstances of its discovery."[8]

In 1947 Kenneth Oakley of the British Museum of Natural History Geology Department suggested that a fluorine test be applied to the Piltdown fossils. It had been known for some time that fossil bones and teeth gradually absorb fluorine from the environment in which they are buried, and that this characteristic can be used to find the relative ages of fossil materials that are buried in the same deposit. The test cannot determine absolute ages in years because the fluorine content of soil, gravel, etc. varies from place to place.

Oakley worked with J. S. Weiner of the Oxford University Department of Anatomy, and ran fluorine analyses on small samples from the cranium and jaw of Piltdown Man and on the other fossils found at Piltdown. The results indicated that the mastodon and elephant fossils were the oldest, with 2-3 percent fluorine, that the deer and beaver were younger, at 1.6 percent, and that the Piltdown cranium and jaw were much younger still, at 0.2 percent. The results were announced publicly in 1948. So Piltdown Man was not as old as Charles Dawson and the British Museum anthropologists had hoped and claimed. Britain had lost its claim to a representative of the really ancient hominids. Alvan T. Marston wrote:

"It is evident that many mistakes have been made concerning Piltdown Man - many mistakes by many highly qualified and highly placed men. To err is human and none of these men have been divine."[9]

However, that fluorine test did not show a difference between the fluorine content of the cranium and that of the jaw, so the claim that the cranium and jaw belonged to a single individual continued to be defended.

In 1953 Wilfred Le Gros Clark of the Oxford Department of Anatomy joined Oakley and Weiner to perform further fluorine tests on the Piltdown cranium and skull, using larger samples than the earlier test. They found the cranium fragments from Piltdown to contain 0.1 percent fluorine, and the jaw 0.03 percent. So the jaw and cranium didn't belong together, either. The results were announced in the Bulletin of the British Museum of Natural History in 1953.[10]

A full-fledged reexamination of all the Piltdown fossils followed. Dawson had soaked the cranium pieces and jaw in a solution of potassium permanganate in an effort to preserve the fragile material, with a result that they carried a brown stain. Chemical analyses showed that other chemicals had also been used to give a stain that gave the impression of having lain long in the gravel pit. Even the flint tools were found to have been stained on the surface, with no penetration to the interior. File marks on the molar teeth of the jaw indicated that the teeth had been filed down to give the impression of natural wear. The single canine was found to have been filed to make it smaller. Apparently the flints, the other animal fossils, and the cranium pieces and jaw had all been planted in the Piltdown gravel pit.

Who did it? Conjectures have been published, but, to this day, no one knows. Most of the men responsible for the find and the reconstruction at the British Museum had died by the time the fraud was exposed. No one has admitted being the hoaxer, and no certain evidence has been found to identify the perpetrator.

Why would someone, anyone, do such a thing? Was it a practical joke that got out of hand? Was some person or persons so eager for fame as to attempt such an elaborate hoax?

How could so many people, experts in their fields and highly placed, be deceived for such a long time?

Conjectures about the hoaxer, and analyses of the case, have been published at book length. Read the books, found in any library, if you are interested in more information and attempts at explanation.

The Piltdown Man case is undoubtedly one of those occasions when the investigators were able to "see what they wanted to see." All of us - professional scientists, people looking for a quick profit from investment opportunities and becoming scam victims, lonely people looking for friends and finding predators who steal their wealth, and more - are sometimes too gullible, and we find ourselves inclined to see what we want or expect to see, while not even noticing evidence that contradicts our perception. Promoters of the view that Earth is a recent creation are as vulnerable to that failing as others, and there are an ample number of cases in which that appears to have been a factor in publishing faulty claims of scientific support for their cause.

In the course of time, as noted above, the Piltdown hoax was discovered and exposed. Scientific endeavor, the study of God's world, has more safeguards than most vocations for protecting against deceiving ourselves. There is usually total openness to the data, so that anyone can examine the explanations that are offered to see if they are valid. As noted above also, those who dissent from a proposed explanation or idea are free to publish their dissent and the reasons for it. The publicity that has accompanied attempts at deceit, such as the "painted mouse" affair and the Piltdown hoax, makes us acutely aware of the hazards of "seeing what we want to see." Since none of us is perfect, we need to be vigilant, keeping our eyes and minds open to evidence for and against the ideas that come our way.

The "young-Earth" perspective

The Piltdown Man incident is mentioned in many publications that promote the view that Earth is a recent creation. The usual context, and sometimes the direct accusation, is that science is unreliable, and that the results of scientific study are not to be trusted, especially when the topic is the history of humans.

For example, in summarizing the lessons to be learned from the

Piltdown hoax:

> "In any of the sciences, and particularly the discipline of anthropology, we might well ask that if men the caliber of Sir Arthur Keith, Sir Arthur Smith Woodward, and Sir Grafton Elliot Smith could have been deceived by their own preconceptions, how can we be sure that men of science today are not also being deceived, not necessarily by hoax, but by their own expectations? After all, those same preconceptions are still very much in the minds of the Leakeys, Johansons, and others involved in the great African fossil hunt today."[11]

And, in concluding a discussion proclaiming doubts about the authenticity of claims of human-like characteristics of *Sinanthropus* (Peking Man):

> "We believe at the very least a combination of prejudice, preconceived ideas, and a zeal for fame have been responsible for elevating a monkey-like creature to the status of an ape-like man. The same combination that produced Peking Man also produced Nebraska Man from a pig's tooth, Piltdown Man from a modern ape's jaw, and Leakey's East Africa Man from an australopithecine."[12]

And, in noting the 50th anniversary of the exposure of the Piltdown Man fraud:

> "In fact, historical revisionists claim that the Piltdown hoax is evidence of the *strength* of evolutionary 'science' because it is self-correcting - given time, scientists will weed out mistakes and errors. But the real question is: why did it take so long for scientists to discover such an obvious fraud? The obvious answer is that scientists are so blinded by their faith in human evolution and 'missing links' that they 'see' things that just aren't there."[13]

Evaluation

Everyone agrees that it took longer than it should have for the forgery to be discovered. The secrecy and possessiveness with which the samples were held by the British Museum were factors; very few people were permitted to have a look at the skull and jaw bone. The desire on the part of the British scientists to have a British sample of an ancient hominid probably played a part. But the really important question is: Is the science of paleoanthropology more reliable now than it was before the forgery and its discovery, or is it less reliable?

Perhaps asking a few questions, and finding the answers to those questions, will help us to decide:

1. Who was responsible for exposing the fraud?

The answer is: scientists in anthropology and related fields, including some connected with the British Museum of Natural History.

2. How was the hoax discovered?

The answer is: by applying the techniques developed by scientists in anthropology and related fields.

3. Who published the information confirming that Piltdown was a hoax?

The answer is: The British Museum of Natural History, the very same institution that had been involved in defending the authenticity of Piltdown at an earlier time.

It is certainly true that every new claim resulting from scientific study should be examined with meticulous care to make sure that it is valid, and scientists themselves are sometimes less than open and modest in making and defending their claims. The process is sometimes slow, as the Piltdown affair demonstrates, but the outcome of the Piltdown matter gives us, the general public, reason to believe that, in the long run, scientific investigation leads to reliable results. Historically, science has been more successful than most endeavors at being self-correcting.

The adage "The wheels of the gods grind slow, but they grind exceedingly fine" may be applied to the matters of scientific investigation as well as to the expectation of justice being meted out to the evildoer.

The whole incident is certainly regrettable. If truth be known, however, the Piltdown Man affair is an example of the success of science in discovering and exposing both error and attempts at fraud. The practitioners of scientific investigation, by and large, are honest and truthful; if they are not, they are likely to be found out.

The "Shooting the Messenger" Syndrome

When we have been deceived in one way or another, we tend to react in anger and disappointment. Sometimes our anger may be misdirected. I remember an instructive story about such a case; maybe it was a movie shown on television, or maybe told elsewhere.

The story goes like this: A wealthy and lonely middle-aged woman was befriended by a handsome man somewhat younger than she. He courted her and persuaded her that he loved her and wanted nothing more than to bring her happiness. She came to trust him, and he managed to get his hands on a great deal of her money. Meanwhile, he was cavorting with other women, and made it clear in the rest of his life that he had no intention of marrying the wealthy woman he had befriended; he was only using her to get her money. But, in her loneliness, the woman was convinced that he cared for her, and she trusted him.

A faithful household maid who was employed by this wealthy woman learned a bit about what was going on. The maid was genuinely fond of her employer, and was distressed by her suspicion that the lady was being deceived and robbed by her supposed lover. The maid undertook a more detailed investigation, and collected photographs and other evidence that left no doubt about the man's deception and greed. When she had organized the evidence she had collected, she brought it to the attention of her employer, wanting to alert her to the deception and protect her from having her money stolen.

The wealthy woman looked at the evidence her maid had presented. There was no doubt about the man's deception. After examining all the evidence, she stood for a few moments considering what to do. Then she left the room, returned with a gun in her hand, and shot the maid.

The story illustrates a common tendency among us. We often want to kill the messenger when we are confronted with a message we don't want to believe.

The perpetrator of the Piltdown fraud has not been positively identified. But, clearly, it is not science who is the culprit. The application of scientific techniques and scientific investigation exposed the fraud, and the results of those investigations were first published in the scientific journals and reports.

If truth be known, the publications that promote the view that Earth is a recent creation, and then try to discredit science, especially anthropology, because of its failure to discover and expose the Piltdown hoax quickly, have fallen into the trap of shooting the messenger.

References

[1] Alexander Kohn, *False Prophets*, (Oxford, UK and Cambridge, MA, USA: Basil Blackwell, 1986).

[2] Kohn, *Prophets*, 77.

[3] Bernard Dixon, "When it smells, hold your nose," *The Scientist* 1, No. 2 (17 November 1986), 13.

[4] Charles Blinderman, *The Piltdown Inquest*. (Buffalo, NY: Prometheus Books, 1986).

[5] Ian T. Taylor, *In the Minds of Men*. (sub-title "Darwin and the New World Order") (Toronto: TFE Publishing, 1984), 225-9.

[6] Website www.answersingenesis.org

[7] Franz Weidenreich, *Apes, Giants and Man*. (Chicago: University of Chicago Press, 1943).

[8] Roy Chapman Andrews, *Meet Your Ancestors*. (New York: Viking Press, 1945), 125.

[9] Alvan T. Marston, 1950. "The Relative Ages of the Swanscombe and Piltdown Skulls, with Special Reference to the Results of the

Fluorine Estimation Test." *British Dental Journal*, 88 (June 2, 1950), 292-9. [Quoted in Blinderman, *Piltdown*, 66.]

[9]J.S. Weiner, K.P. Oakley, and Wilfred Le Gros Clark, "The Solution of the Piltdown Problem." *Bulletin of the British Museum (Natural History)*, Geology Series 2, No. 3. (London: British Museum, 1953), 141-6.

[10]Taylor, *Minds*, 229.

[11]Duane T. Gish, *Evolution – The Fossils Say NO!* (San Diego: Creation-Life Publishers, 1973), 103.

[12]website www.answersingenesis.org, (June 5, 2009).

12 Nebraska Man

None of us humans is perfect. Everybody makes mistakes. Since scientists are human, they sometimes make mistakes. Of course, the same is true for philosophers and theologians.

The episode usually referred to as "Nebraska Man" is a widely publicized case of an error in science. The story has been repeated often in publications and lectures by those who disparage the published results of science regarding the ages of rocks and the long history of living organisms on Earth. The story is told in such publications and lectures with the implication, and sometimes the direct accusation, that science is unreliable, and that scientists are not worthy of our trust.

So, we should provide a summary of the history of this error in science, so that you, the reader, may evaluate the integrity of the natural sciences and scientists, as they are engaged in the study of God's world.

Henry Fairfield Osborn, renowned paleontologist at the American Museum of Natural History in New York City, had received a single tooth from consulting geologist Harold J. Cook, who had discovered the tooth in western Nebraska. In an article in the professional journal *Science* dated May 5, 1922, Osborn described the tooth as "a single small water-worn tooth," and proceeded to describe his examination and conclusion. After conferring with an associate, Dr. William D. Matthew, Osborn concluded that the tooth was from an anthropoid ape. He also had the tooth examined by Curator William K. Gregory and by Dr. Milo Hellman, who, Osborn reported, "have both made a special study of the collections of human and anthropoid teeth in the American Museum and the United States National Museum." Osborn published their conclusions in full, and then agreed with them that the tooth had "its nearest resemblances

103

with … men rather than with apes," and he named the specimen *Hesperopithecus haroldcookii*.[1] (The name *Hesperopithecus* means "ape of the west.")

This report was sensational news, since no ancient hominid fossils had been found previously in the Americas. The men who had examined the tooth and identified it as hominid were paleontologists with international reputation. The report received wide distribution and gave rise to a sensationalized article in *The Illustrated London News*, including a diorama of a hominid in the midst of a collection of mammals whose fossils were (and are) found in the same vicinity. The public press of that era was no less eager to seize and sensationalize news items than it is today.

A more extensive article was published in the Museum publication *The American Museum Novitates*, in which some of Osborn's associates published further comments on the Harold Cook tooth, including detailed comparative figures and measurements.[2] Photographs of the Harold Cook tooth were included, along with a molar tooth of a modern native American for comparison; to my unpracticed eye the teeth appear to be very similar.

Some five years later another article about that tooth and its identification appeared in the December 16, 1927 issue of *Science*, the same journal that had published the original announcement in 1922. The article was published over the name of William K. Gregory, the Curator who had examined the tooth 5 years earlier and identified it as hominid. He related the history of the tooth in some detail, and I will quote from that article at some length:

"In the type specimen [that is, the tooth received from Harold Cook] the crown of the tooth had been ground off by long wear to such a degree that the surface of the crown was entirely gone and only the very basal portion was left. This presented an evenly concave surface of wear that was strikingly similar to the worn-down surface of one of the upper molar teeth that had been found by Dr. Dubois at Trinil, Java, near the famous skull top of *Pithecanthropus erectus*. The Nebraska tooth also had a very wide root on the inner side, which was similar to the wide root on the inner side of the upper molars of *Pithecanthropus* and of many teeth of American Indians. Hence Drs. Gregory and Hellman, whose report was cited by Professor Osborn, were inclined to think that on the whole the nearest resemblances of the specimen were with men rather than with apes."

A little later in the report, Gregory reported:

"The scientific world, however, was far from accepting without further evidence the validity of Professor Osborn's conclusion that the fossil tooth from Nebraska represented either a human or an anthropoid tooth. Many authorities made the objection 'Not proven,' which is raised to nearly every striking new discovery or theory, and in course of time nine suggestions were put forward by responsible critics as to what the type specimen of *Hesperopithecus* might represent other than any kind of ape or man."

Later still in the report Gregory continued:

"In the hope of discovering more remains of this highly interesting fossil, Professor Osborn sent Mr. Albert Thomson, of the Museum staff, to collect in the Snake Creek beds of Nebraska in the summers of 1925 and 1926. ... Among other material the expedition secured a series of specimens which have led the writer [Gregory] to doubt his former identification of the type as the upper molar of an extinct primate, and to suspect (it) to be an upper premolar of a species of *Prosthennops*, an extinct genus related to the modern peccaries."

And later:

"The still weak link in the chain of evidence consists in the fact that in *Prosthennops* the premolars that approach the type tooth of *Hesperopithecus haroldcookii* have two inner roots, whereas the type tooth has a single broad root."

In a "POSTSCRIPT" appended to the main article Gregory reported:

"Last summer (1927) Mr. Thomson made further excavations in the exact locality where the tooth (found by Harold Cook) was discovered. A number of scattered upper and lower premolar and molar teeth were found in different spots, but every one of them appears to me to pertain to *Prosthennops*, and some of these also resemble the type of *Hesperopithecus*, except that the crown is less worn. Thus it seems to me far more probable that we were formerly deceived by the resemblances of the much worn type to equally worn chimpanzee molars than that the type is really a unique token of the presence of anthropoids in North America."[3]

Signed by William K. Gregory.

And in this way the error of mistaken identification was corrected.

Initially a mistaken identification, to be sure. But an honest mistake, with reasons given for the original identification on the basis of similarities to other hominid specimens. The error was admitted and corrected in a widely distributed scientific journal by one of the paleontologists who played a major role in the initial study and identification of the tooth.

In addition, the comment by Gregory above "Many authorities made the objection 'Not proven'" demonstrates that there was by no means universal agreement by other competent scientists with the identification published by Osborn at the time of the first publication of that mistaken identification.

The response in publications that promote the view that Earth is a recent creation

Now let us note how the story is told in some of the literature promoting the view that Earth is a recent creation. The story is told in detail in a book entitled *In the Minds of Men*.[4] I will quote a passage on that subject from a later article by the same author entitled "Nebraska Man revisited," as follows:

"More about 'Nebraska man' - that now-discarded pig's tooth that was reconstructed by some to look as though it came from a primitive evolutionary 'ape-man'.

"When Dr Henry Fairfield Osborn, head of the Department of Paleontology at New York's American Museum of Natural History received the fossil tooth in February 1922, he would have thought it a gift from the gods had he believed in any god at all. Marxist in his views and prominent member of the American Civil Liberties Union, he was aware that plans were being made by the union to challenge the Christian-backed legislation that forbade the teaching of evolution in American schools. He saw the tooth as precious evidence for the test case which was eventually held in 1925 at Dayton, Tennessee, and became known as the 'Scopes Monkey Trial'.

"The trial was an arranged affair, but the tooth was not brought in as evidence because there was dissension. The truth leaked out slowly and obscurely in the *American Museum Novitiates*[sic] for January 6, 1923, where nine authorities cited their objections to the claim that the tooth was anywhere near related to the primate. A further search was made at Snake Creek, the site of the original discovery, and by 1927 it was begrudgingly concluded that the tooth was that of a species of *Prosthennops*, an extinct genus related to the modern peccary or wild pig. These facts were not considered generally newsworthy but did appear in *Science* (1927, **66**:579). The fourteenth edition of the *Encyclopaedia Britannica* (1929, **14**:767) coyly admitted that a mistake had been made and that the tooth belonged to a 'being of another order'. The burden of embarrassment was thus eased for the now retired Henry Fairfield Osborn."[5]

This is one of many instances in which science and scientists have been the object of ridicule in lectures and publications promoting the view that Earth is a recent creation. Let us note some of the language that is used in the examples quoted above:

1. The tooth in question is referred to here and in other publications that promote the view that Earth is a recent creation as a "pig's" tooth. But the tooth is actually that of a peccary, and a peccary is not a pig. Not only are the peccary and the pig not the same species in the Linnaean classification scheme, but they are not

even in the same genus, nor even in the same family. But "pig" has a more loathsome ring to it, don't you think?

2. The scientists who were guilty of the misidentification are usually referred to in publications that promote the view that Earth is a recent creation in terms that suggest an unwillingness to recognize or admit their errors. Note the terms "reconstructed by some to look as though" and "The truth leaked out slowly and obscurely" and "it was begrudgingly concluded" and [Encyclopaedia Britannica] "coyly admitted."

The truth of the matter

What is the purpose of language that imputes subversive motives to publications whose language is entirely sober and straightforward?

It appears that the purpose of such language in the oft-repeated story of an error in science is to cast doubt on the integrity of the scientists involved, and to cast doubt on the reliability of the results of the science being pursued.

But does the actual story reflect a lack of integrity on the part of the scientists, or a lack of reliability on the part of the science? I think not. Rather, the episode demonstrates a high level of integrity and reliability of scientific investigation. Note the sequence: 1) a tooth is found and submitted to an eminent authority for identification, 2) an identification is published, with supporting evidence and documentation, 3) other competent scientists examine the specimen and the evidence, 4) doubts about the validity of the initial identification are published by other scientists, 5) further investigation is undertaken - Osborn himself sent an assistant to make further collections in the same area where the initial tooth was found, 6) the additional evidence is examined and found to refute the initial identification, and 7) the results of the re-examination of the evidence are published, admitting and correcting the initial error.

I find no "coyness" or "begrudgingness" or attempts to evade the admission of error in the quoted passage from *Science* given above; do you? It is an indication of integrity when we can admit that we have made a mistake. It is especially a sign of integrity when we recognize and correct *our own* mistakes.

It is also worth noting that the initial announcement of the find, and the mistaken identification of the tooth as probably belonging to a primate, was published in the professional journal *Science*, published by the American Association for the Advancement of Science (AAAS). *Science* was then, and is now, the most widely distributed professional scientific journal in North America. The admission of error and the correction were also published in *Science*, the same journal that carried the initial, mistaken identification. The comment in the article by Taylor, above, that "the truth leaked out slowly and obscurely" is far from the truth.

Additional comments

It is interesting to note that arguments leveled against the reliability of the results of scientific investigation in publications that promote the view that Earth is a recent creation are very selective in the results that are objected to. We readily accept results that bring better medical care and higher yield farm crops and more creature comforts; why do some of us choose to ridicule the reliability of science on only certain selected topics?

There is another matter related to this discussion which we should note. I am acquainted with some of the proponents of the view that Earth is a recent creation, and I know that they are committed Christians. As Christians, we are fond of singing "They'll know we are Christians by our love." So, is there an appropriate place in the life and speech of a Christian for sarcasm and ridicule of persons? I think not. Yet, comments about errors in science and mistakes by scientists found in publications and lectures that promote the view that Earth is a recent creation are often replete with sarcasm and ridicule of persons; do you think that is OK?

Everyone makes errors now and then. I have, you have, the people who misidentified the tooth that became known as "Nebraska Man" did. Hopefully the old adage "We benefit from our mistakes; we recognize them as mistakes when we make them again" is not entirely true. Hopefully we learn not to make the same ones again.

The science of paleontology is carried on by humans, and it is therefore always imperfect and incomplete. But, as the full story of "Nebraska Man" demonstrates, the science of paleontology works. Errors are discovered, admitted or exposed, and corrected.

References

[1] Henry F. Osborn, "*Hesperopithecus*, the first anthropoid primate found in America," *Science* 55 (May 5, 1922), 463.

[2] William K. Gregory, Milo Hellman, and William D. Matthew, "Notes on the type of *Hesperopithecus haroldcookii* Osborn," *American Museum Novitates*, No. 53 (January 6, 1923), 1-16.

[3] Gregory, "*Hesperopithecus* apparently not an ape nor a man," *Science* 66 (December 16, 1927), 579.

[4] Ian T. Taylor, *In the Minds of Men*. (sub-title "Darwin and the New World Order") (Toronto: TFE Publishing, 1984), 231-3.

[5] Taylor, "Nebraska man revisited," *Creation Ex Nihilo* 13, No. 4 (1991), 13.

13 The Second Law of Thermodynamics

Wow! That sounds like it might be complicated! But we should deal with it, because it is sometimes touted as being very important with regard to biological evolution by natural processes, and so it looms large in some of the publications that promote the view that Earth is a recent creation.

OK. We'll put the more complicated, more mathematical stuff in Appendix A for those readers who want to pursue the topic in greater detail, and we'll present the basic ideas here in simple prose.

The word "thermo-" "dynamics" refers to "heat" and "motion," or "heat" and "energy." The behavior of heat and energy in our physical world was of considerable scientific interest during the 19th Century in connection with the development of steam power for driving ships, trains, and electrical generators. Much of our present understanding of thermodynamics was developed as a result of that interest. Our understanding is expressed in ideas, or explanations, that are called the "Laws of Thermodynamics."

The First Law of Thermodynamics

The "First Law of Thermodynamics" has more commonly been called "The Law of the Conservation of Energy," and simply says that energy is conserved in any physical or chemical process. The form of energy may be changed: changing energy of motion into heat energy, as, for example, in applying the brakes in a moving automobile; or changing electrical energy into light energy in an incandescent light bulb (actually, part of the electrical energy is converted into light and part into heat); or changing heat energy into energy of motion in a steam engine; etc. But there is no loss or gain of energy in making the conversion from one form of energy into another. "Energy is neither created nor destroyed in a physical or chemical process."

We should note that our present understanding recognizes that matter can be converted to energy is some circumstances, as, for example, in a nuclear reactor.

The Second Law of Thermodynamics

The "Second Law of Thermodynamics" introduces a new concept called "entropy." Entropy is a measure of the state of order or "orderedness" of some portion, or all, of the universe. The sense of "order" under consideration here has to do with the ordered arrangement of atoms and molecules in a material substance, such as the geometric arrangement of atoms in a mineral crystal. (The Second Law of Thermodynamics has nothing whatever to say about the state of a teenager's bedroom; the proper terminology for that situation would be "orderly," or, in some cases, "disorderly.")

A physical or chemical process may result in a change in the entropy of the material involved in the process. An increase in entropy amounts to a decrease in orderedness; a decrease in entropy amounts to an increase in orderedness. In solid water (ice) the water molecules are found in an ordered geometric arrangement; in liquid water the molecules have a higher entropy, that is, they can move about within the liquid, they are not in a fixed order, but they are still limited by the surface of the liquid; in gaseous water (steam, water vapor) the molecules have a higher entropy still, that is, they have even less order than in the liquid, and they can move about more freely, within whatever container encloses the water.

A classical way of stating the Second Law of Thermodynamics is as follows: "The entropy of the universe tends toward a maximum." Stated in somewhat more detail, the entropy of the universe never decreases as a result of any spontaneous physical or chemical process. In some processes the entropy of the universe may remain unchanged, and in others it may increase, but it never decreases. It is important to note that this statement has to do with the entropy of **the universe** as a whole; we will have some things to say later about local "systems," or local regions within the universe.

After many decades of study, the Second Law of Thermodynamics is considered to be universal, that is, it applies to all physical and chemical processes. No exception has ever been observed.

A spontaneous process is one that takes place irreversibly, that is, the interaction proceeds from one state or condition to a different state or condition, but never goes spontaneously in the opposite direction. There are many examples: the burning of natural gas; the oxidation of iron in the presence of oxygen (rusting), etc. Such spontaneous processes always result in an increase in the entropy of the universe.

In carrying out experiments, or in conducting business and manufacturing, or in observing natural processes, we are generally not dealing with the entire universe (We wouldn't know how to do that!), but with only a small component or region within the universe. We speak of such a restricted region or component as a "system" for our attention or study.

For the laboratory study of the thermodynamic properties of various materials and the changes that occur in various processes, the experiment is carried out in a well-insulated container, called a "calorimeter," so that the measurements can be done without any heat or other energy being lost to or gained from the surroundings. We call such a system a "closed" system.

Is there such a thing as a totally "closed" system, with absolutely every interaction with its surroundings accounted for by the experimenter? Probably not, but there is reason to think that apparatus can be built that comes very close to that ideal.

Most of the processes that are of interest to us, however, are "open" systems, in which processes are taking place under conditions that do not isolate the local system from its surroundings. When we try to learn about changes in entropy within such a system, we must remember that we will have to examine <u>both</u> the restricted system itself, <u>and</u> its surroundings, in order to account for what is happening to the entropy of the universe as a whole. Methods that have been developed for doing such an examination are described in Appendix A.

We have provided a general definition of entropy in prose, but entropy can also be defined mathematically. As promised, the mathematical treatment is found in Appendix A. However, since we find some claims in publications that promote the view that Earth is a recent creation that are concerned with the Second Law of Thermodynamics, we will comment on those here.

The perspective in some publications that promote the view that Earth is a recent creation

Some publications that promote the view that Earth is a recent creation have made the claim that biological evolution by natural processes is impossible, because such evolution would be contradictory to the Second Law of Thermodynamics.[1,2,3] That claim continues to be published; it can be found on the website www.christiananswers.net, 2009. Since biological evolution produces an increase in order, so it is claimed, such evolution could not occur by spontaneous natural processes, since the Second Law of Thermodynamics states that all spontaneous processes result in a decrease in order (an increase in entropy).

The discussion in the book *Scientific Creationism*, in a chapter entitled "UPHILL OR DOWNHILL?" expresses the claim in the following words:

"The Second Law of Thermodynamics is particularly important in this discussion, since it states that there exists a universal principle of change in nature which is downhill, not uphill as evolution requires."[4]

and

"We are warranted, then, in concluding that the evolutionary process (the hypothetical Principle of Naturalistic Innovation and Integration) is completely precluded by the Second Law of Thermodynamics."[5]

If truth be known, however, the claim is mistaken.

There are numerous processes taking place spontaneously on Earth every day in which there is an increase in orderedness (decrease in entropy) in the local system. Such local systems are not isolated from their surroundings, of course. In such processes, the decrease in entropy in the local system is accompanied by a greater increase in the entropy of the surroundings, so that there is an increase in the entropy of the universe as a whole. Therefore, there is no violation of the Second Law of Thermodynamics in such processes.

Here are a few examples of processes that occur spontaneously, and that result in a <u>decrease</u> in the entropy of the local system: 1) the burning of natural gas in air or oxygen to produce carbon dioxide and water; 2) the burning of magnesium metal in air or oxygen to form magnesium oxide (as is common in fireworks); 3) the burning of hydrogen gas in air or oxygen to form water; 4) the oxidation (rusting) of iron in the presence of air or oxygen (yes, even the rusting of iron, so frequently used as symbolic of the "decay and deterioration" of our world, results in a decrease in the entropy of the local, open system).

In each of these cases, and in every open system, the surroundings, the rest of the universe, act as a large constant-temperature heat sink or heat source. In each of these cases, the heat or energy that is absorbed or supplied by the surroundings produces an increase in the entropy of the universe that is greater than the decrease in entropy of the local system. The relationship of an open system to its surroundings is treated in greater detail in Appendix A.

Many other examples could be given, but these suffice to demonstrate that local, open systems can experience a <u>decrease</u> in entropy in spontaneous, irreversible processes. If truth be known, the claim that a decrease in the entropy of all local, open systems by natural, spontaneous processes would be contradictory to the Second Law of Thermodynamics is mistaken.

We should note here that the website www.answersingenesis.org has published a list of claims that are known to be faulty, and which should not be used to defend the view that Earth is a recent creation; that list includes the claim that the Second Law of Thermodynamics was placed into effect only after the disobedience of Adam and Eve. However, the list of faulty claims on that website does not include any comment about the compatibility of the Second Law with biological evolution.

Biological systems and the Second Law of Thermodynamics

And what about biological systems?

First, there is a question about the meaning of "order" in the language of the physicist/chemist in discussing entropy, and the meaning of "order" in the language of the biologist/taxonomist.

As mentioned earlier, the thermodynamic meaning of "order" has to do with the ordered arrangement of atoms and molecules in a material system. If a local system has undergone a process that results in a greater order, in that thermodynamic sense, it has undergone a decrease in entropy.

The biologist, speaking of one organism or species as being of higher "order" than another is referring to the traditional classification of living organisms, or taxonomy. The characteristics that determine "higher" or "lower" order, in that sense, do not depend on differences in thermodynamic entropy states.

Biological evolution is commonly thought of as progressing from lower order to higher order. What is meant by such a concept? Is the brain cell of a "higher order" primate in a more highly ordered state, in the thermodynamic sense, than the nerve ganglion cell of an earthworm?

It is not obvious that the Second Law of Thermodynamics has anything at all to do with biological evolution, that is, with the transition from one species of organism to another by whatever natural and spontaneous processes are involved. However, if biological evolution occurs on Earth, including processes that result in an increase in order in the thermodynamic sense, and thus a decrease in entropy, those processes are occurring in local, open systems. The Second Law of Thermodynamics certainly does **not** preclude the occurrence of a decrease in the entropy of local, open systems.

Concluding Comment

The mistaken claim in *Scientific Creationism* must have been based on a misunderstanding and misapplication of the Second Law of Thermodynamics, that is, in using the Second Law of Thermodynamics as it pertains to the universe as a whole and applying it to the local, open system <u>without</u> considering what happens to the entropy of the surroundings.

How could this claim - the claim that <u>all</u> spontaneous processes result in an <u>increase</u> in the entropy of open, local systems on Earth - have found its way into the publications that promote the view that Earth is a recent creation, since it is so obviously mistaken? As is demonstrated in Appendix A, every college freshman chemistry textbook published over the past many decades includes a discussion of the Second Law of Thermodynamics as applied to chemical reactions in local, open systems, as well as to the universe as a whole, and exposes the error of the claims presented in *Scientific Creationism* and other publications that present the same faulty claim.

(See Appendix A for the more detailed, technical discussion.)

References

[1]Bolton Davidheiser, *Evolution and Christian Faith.* (Nutley, New Jersey: Presbyterian and Reformed Publishing Company, 1969), 220-4.

[2]Paul D. Ackerman, *It's A Young World After All.* (Grand Rapids: Baker Book House, 1986), 113.

[3]Henry M. Morris, Ed., *Scientific Creationism.* (San Diego: Creation-Life Publishers, 1974), 37-46.

[4]Morris, *Creationism,* 38.

[5]Morris, *Creationism*, 45.

14 Thermodynamics II - Error upon Error

Biological systems and biological growth

The growth and development of an individual living organism takes place by chemical processes in which complex molecules are constructed from simpler molecules, involving an increase in orderedness, and thus a decrease in entropy. This takes place by spontaneous, natural means. In the biological realm, then, as well as in the non-living cases for which examples were given earlier, there are many known cases in which the entropy of a local, open system decreases as the result of spontaneous, natural processes.

The growth of a living organism, then, presents a serious challenge to the claim in publications that promote the view that Earth is a recent creation that any decrease in the entropy of a biological system would be contradictory to the Second Law of Thermodynamics.[1] Because growth of living organisms so obviously occurs by natural, spontaneous processes, there is some discussion in the book *Scientific Creationism* concerning the growth of living organisms and the Second Law. The claim is made that:

"There do exist a few types of systems in the world where one sees an apparent increase in order, superficially offsetting the decay tendency specified by the Second Law. Examples are the growth of a seed into a tree, the growth of a fetus into an adult animal, and the growth of a pile of bricks and girders into a building."[2]

[**NOTE (by CM):** The growth of a pile of bricks and girders into a building does **not** take place by spontaneous, natural processes, so this example doesn't apply to the question at hand, and will be ignored in this discussion.]

The quotation continues:

"Now, if one examines closely all such systems to see what it is that enables them to supersede the Second Law .... he will find in every case, at least two essential criteria that must be satisfied:

(a) *There must be a program to direct the growth.*

".... In the case of the [living] organism, this is the intricately complex genetic program, structured as an information system into the DNA molecule for the particular organism. ....

(b) *There must be a power converter to energize the growth.*"[3]

So, a few comments about that claim:

1. The Second Law of Thermodynamics is considered to be a "universal" principle, and is described as such in *Scientific Creationism*. Since it is a universal principle it cannot be "superseded" under any circumstances whatever; if it were sometimes superseded, it could not be characterized as "universal."

2. In addition to carrying genetic information from one generation to the next, the DNA in a living organism does direct the growth in a sense, that is, it provides the template for the formation of proteins, but the individual chemical reactions that take place in the growth process occur entirely in accord with the Second Law of Thermodynamics; it is a universal principle, after all. As in the non-biological examples given earlier, the natural, spontaneous chemical reactions taking place in the growth process may occur with a decrease in entropy in the local, open system, while the entropy of the universe increases because of the greater increase in the entropy of the surroundings, as we noted in the previous chapter.

3. The claim in *Scientific Creationism* speaks of "an apparent increase in order, superficially offsetting the decay tendency specified by the Second Law." However, the increase in order that is seen is very real, not merely "apparent." And there is nothing superficial about the changes that occur in the growth of a living organism.

Conclusion

If truth be known, the growth of a living organism by natural, spontaneous processes, with its increase in orderedness and decrease in entropy, takes place without any violation or "superseding" of the Second Law of Thermodynamics.

The notion of "superseding" the Second Law was apparently introduced in the publications that promote the view that Earth is a recent creation in an attempt to reconcile the obvious decrease in entropy in the growth of living organisms with the claim that the Second Law precludes a decrease in entropy in **all** local, open systems by natural, spontaneous processes. If truth be known, both that notion and that claim are mistaken.

All attempts to defend error beget further error, usually leading to internal inconsistency, as in this case.

References

[1] Henry M. Morris, Ed., *Scientific Creationism*, (San Diego, Creation-Life Publishers, 1974), 37-46.
[2] Morris, *Creationism*, 43.
[3] Morris, *Creationism*, 44.

15 An Interlude - Columbia Basin Lava Flows

Well, for several chapters we've been considering mistaken and unworthy claims that supposedly support the idea that the Earth is not more than several thousand years old. So let's take a break from that stuff, and take a few pages to look at God's Earth from a positive slant. Let me tell a true story of a surprising discovery.

The bedrock in central Washington consists of dark-colored lava, deposited from volcanic activity of the past. A steep-sided channel, known as the Grand Coulee, cuts through the lava flows from north to south. There are several small lakes on the floor of the channel. The surprising discovery was made along the shore of Blue Lake, a few miles south of Dry Falls State Park.

In the summer of 1935 two couples from Seattle were hiking along the cliffs bordering the northeast arm of Blue Lake. The cliffs are in lava flows known as the *Columbia Basin Basalts*. The hikers came across a large cavity in the basalt, and one of them crawled inside. At the far end of the cavity he found part of a jawbone with teeth in it. The bones and teeth were given to the University of Washington by one of the hikers.

Professor George F. Beck of Central Washington College of Education at Ellensburg heard about the discovery, and visited the cave with an assistant. Additional bones were collected from extensions along the side of the cavity. These bones and the fragments of jawbone and teeth from the University of Washington were identified by Dr Chester Stock of the California Institute of Technology as belonging to a rhinoceros. The find was reported in the scientific literature.[1,2]

In 1948 J.W. Durham and D.E. Savage of the University of California at Berkeley visited the site and undertook a very careful examination of the cavity, working with W.M. Chappell, who had visited the site in 1937. Additional broken fragments of bone were found, including some from the left front leg of the animal. Casts were made of sections of the cavity, using burlap and plaster, and the casts were reassembled to make a single cast of the entire cavity. This cast conforms very nearly to the shape of living species of rhinoceroses. The rhinoceros was lying on its left side, with its legs extended upward. The legs are spread apart, as would be the case for an animal that had become bloated with early decay a couple of days after dying. This cast is now in the University of Washington Burke Museum in Seattle.[3]

Comparison of the bones, primarily the jaw and teeth, with other fossil rhinoceroses found in North America indicate a close similarity to *Diceratherium annectens*, a specimen found in the John Day formation in Oregon. The cavity as found in 1935 enclosed nearly all of the animal, except for the rump and most of the left hind leg, which would have extended beyond the present cliff surface. The opening to the cavity was exposed by erosion of lava that would have enclosed the rearmost part of the rhinoceros. The complete animal was nearly eight feet in length from nose to rump.

The rhinoceros mold is at the base of the Priest Rapids basalt flow. The cavity is surrounded by "pillow" basalt, a deposit of rounded, lobe-shaped masses of lava packed together. (Pillow lava is formed when molten lava enters water, as, for example, along the ocean shore on the Big Island of Hawaii.) Pillow lava also extends to the right and left (approximately NNW to SSE) of the cavity at the base of the basalt flow. The thickness of the pillow basalt structure is about two feet to the right and left of the rhinoceros cavity, but thicker at the rhinoceros mold, extending about two feet above the cavity. The pillow basalt grades upward into massive basalt, a much more common form in the *Columbia Basin Basalts*. Below the pillow basalt of the Priest Rapids flow, and above the underlying basalt flow, is a sandy sedimentary layer averaging about five inches in thickness. At several places near the rhinoceros cave, there are fossilized tree stumps and logs resting on the sedimentary layer, and also several cylindrical cavities that appear to be log molds. All of these, like the rhinoceros mold, are surrounded by pillow basalt.

So, what sort of historical sketch can be inferred from the information available to us? The usual story is this: 1) the rhinoceros died and the body was lying in or next to a shallow pond or lake; 2) the lava flow came over the area, encountered the water of the shallow lake, forming the pillow structure typical of hot lava encountering water; 3) the lava covered the body of the rhinoceros, with surfaces cooled enough to be solid, but still flexible enough to conform to the shape of the animal; 4) the pillow lava was chilled quickly enough to support the lava flowing over it without collapsing and crushing the animal's dead body; 5) the entire lava flow solidified, was covered with subsequent sediments and later lava flows; 6) erosion removed the rock adjacent to the rhinoceros mold; 7) the cavity was found by hikers in 1935. A very interesting case of unusual preservation of evidence of living organisms of the past.

Flood Basalts

There are only a few places in the world where we find deposits of lava called "flood basalts." These are black or very dark volcanic rocks that flowed from fissures in Earth's surface and covered a broad region, but without building a cone-shaped mountain around the vent of the volcano. The lava was apparently very fluid, or low viscosity, and flowed over the landscape much like a water flood would.

We have not observed any such lava flows taking place in modern times. However, the rocks formed by such flows display many similarities to modern lava flows, and we can learn something about their history by features that resemble modern lava flows. At the same time, they display some differences from modern lava flows, and we will take note of those.

The largest flood basalt deposit in the world is found in southern India, and is known as the *Deccan Basalts*, or sometimes referred to as the *Deccan Traps*. The second largest such deposit in the world covers 50,000 square miles of portions of Washington, Oregon, and Idaho, and is known as the *Columbia Basin Basalts*. We'll take a more detailed look at the rocks of the Columbia Basin.

Flow structures

The deposit of basaltic lava flows in south central Washington is somewhat more than 10,000 feet thick at its thickest, and consists of about 100 separate flows. The lower flows are depressed downward in the central region of the deposit, which gives rise to the title "basin."

Each flow has a distinct and identifiable lower contact and upper surface. The lower part of the flow, where the molten lava came into contact with the previously existing surface, was cooled rapidly to form solid basaltic rock, often containing a few gas vesicles where bubbles of gas were trapped in the cooling lava. The upper surface, in contact with the air above, cooled even more quickly, trapping many gas vesicles and often displaying the same sort of lava surface flow features that we find at the surface of modern lava flows, in Hawaiian "pahoehoe," for example.

The central part of the flow, which cooled more slowly, typically displays "columnar jointing," a structure of mostly vertical columns, each column usually with four or more sides, formed by the shrinkage of the rock as it continued to cool after it had solidified. Most of the flows consist of two sets of columns; the lower set has thicker columns where the heat escaped by flowing downward, and the upper set has thinner columns where the heat escaped by flowing upwards. One of the flows exposed along the east side of the Columbia River at the I-90 highway bridge has three sets of columns.

The individual flows range in thickness from a few tens of feet to more than 100 feet. Several of the upper flows have been given individual names, and have been placed in the proper sequence in the local geological column.[4] To provide some understanding of the low viscosity of the flow, we note that the Roza Flow, about 80 feet thick, can be traced over an area of 20,000 square miles. The lava must have flowed nearly like water over a nearly horizontal surface to cover such a large area before it cooled sufficiently to congeal and stop the flow. The lava in Hawaii, by comparison, will not flow on a surface that slopes at an angle of less than 3°. Also, the lava of the Roza flow must have been extruded in less than a week to cover such a large area before the lava cooled sufficiently to stop the flow.

In some places where the lower flows have been exposed by erosion, feeder dikes have been found where the molten lava flowed through a fissure in the underlying flow to bring the molten lava to the surface to form the overlying flow. Such fissures are typically only several feet wide, and they extend several miles in length. Each flow was probably brought to the surface through a system of many such fissures.

The greater part of each of the lava flows was extruded, cooled and solidified under air. The lower and upper contacts and the columnar structure that are described above are evidence of that. There are some local areas, however, where the lava has a different structure. In these locations, the lava has the form of "pillow" lava, described earlier, and often with a yellowish mineral called pelagonite at the contacts between pillows. Where we find pillow structure in the *Columbia Basin* flows, the lava apparently encountered water; a pond, a lake, or a stream. A pillow basalt deposit can be observed at the top of the cliff on the east side of the Columbia River at the I-90 highway bridge, and can be observed close up alongside State Highway 26/243 a short distance south from I-90.

Interflow zones

In many places, sedimentary deposits are found between successive lava flows of the *Columbia Basin*. These consist of soils formed by weathering of the upper surface of the underlying flow, and sand and silt carried onto that surface, which then got covered by the overlying flow. These interflow sediments commonly contain fossil plant material, providing evidence of a span of time between successive lava flows.

One of the prominently exposed interflow zones is the Vantage Sandstone, found at the surface just west of the Columbia River along highway I-90 at Vantage, WA. The sandstone layer is found between the Yakima Basalt member (below) and the Frenchman Springs member (above). The Vantage Sandstone contains an abundance of fossilized wood, including large logs of a wide variety of species such as pine and gingko. The Gingko Petrified Forest State Park along I-90 at Vantage displays numerous samples of such petrified logs in place in the sandstone. Amateurs (including this writer) have collected petrified wood from a broad area where the Vantage Sandstone is exposed.

There are other interflow deposits, less prominently exposed than the Vantage Sandstone, perhaps, but very important for understanding the history of the region. Along the Palouse River, upstream from the Palouse Falls at Palouse Falls State Park, there is a footpath that lies along the contact between two basalt flows. (This footpath is not accessible from the State Park; it must be approached from upstream.) Along that footpath, about a quarter mile upstream from the Falls, there is a paleosol, a soil deposit, between the lava flows. The soil contains fossil imprints of wide-leafed grassy plant material, and also some coal and coalified plant material. I have personally seen and collected material from that site.

Well logs

At the western edge of the basalt flows, the rocks have been squeezed and compressed, mostly from north to south, and folded into parallel ridges, mostly trending east-to-west. The folding took place late in the sequence of lava deposits and since then. The area has not been extensively weathered and eroded since the folding of the rock layers, so the up-folded rocks, the anticlines, form the high ridges, and the down-folded rocks, the synclines, lie in the valleys between ridges.

Anticlines are objects of interest for oil companies, because anticlines often form the traps in which crude oil and natural gas may be collected and preserved. True to form, some exploration wells have been drilled in the anticlines of the *Columbia Basin Basalts*.

The Rattlesnake Hills No. 1 Well was drilled for the Standard Oil Co. of California in 1957 and 1958, and was drilled to a depth of 10,655 feet below ground surface. The bottom of the well was still in basalt. While a small amount of natural gas was pumped from the well, the amount did not justify continued pumping, and the well was plugged with concrete from 540 feet to 690 feet below ground surface, and the well was abandoned. In 1967 the well was re-entered and re-drilled for Battelle Pacific Northwest Laboratories for study of the basalt structures.[5]

The drill log for Rattlesnake Hills No. 1 Well shows that 75 separate lava flows were encountered from ground surface to a depth of 10,000 feet, with an equal number of interflow zones. The thicknesses of identified basalt flows ranged from 45 feet to 102 feet; the thicknesses of interflow zones ranged from 14 feet to 37 feet.[6] A considerable amount of coal was encountered in the interflow zones.

Among other studies and measurements done by Battelle Pacific Northwest Laboratories, coal samples from interflow zones were collected and examined for spore and pollen studies (palynology). Thirty-eight taxa (genera) of fern spores, gymnosperm (conifer) pollen, angiosperm (flowering plant) pollen, and algae were identified in the samples.[7] The bulk of the samples represent a temperate-to-warm temperate-to-(possibly) sub-tropical climate. The genera identified were common in the region during the Oligocene Epoch. There is no evidence of arid-type plant pollen in any of the samples, suggesting that the mountains or hills of the region were relatively low, and that there was abundant rainfall (by contrast with the current conditions, in which the *Columbia Basin* lies in the rain shadow of the Cascade Mountain Range).

Another well, the Burlington Northern #1-9 Well, was drilled for Shell/Arco in 1982-84 at the peak of the anticline forming Saddle Mountain north of the Hanford Reservation. The well reached the bottom of the basalt flows at a depth of 12,800 feet below ground surface. The drilling continued to a depth of 17,518 feet, encountering sedimentary deposits, volcanic tuff, and andesitic igneous rock. Numerous layers of coal were encountered in the sedimentary layers. Core samples were collected at various depths; I have had opportunity to examine one section of core from a depth of 15,300 feet. This core consists of fluvial (flowing water) sandstone from within a 60-foot thick sandstone layer, and the core sample contains small globs of clay and strands of coalified plant material.

Two more wells that were drilled through the *Columbia Basin Basalts* deserve mention, both in the Yakima ridges toward the western edge of the lava flows. 1) Shell Bissa #1-29 was drilled in 1982, reached the bottom of the basalt flows at a depth of 4600 feet, then continued through sediments with numerous coal occurrences and into granitic igneous rocks to a depth of 14,965 feet. 2) Shell Yakima Minerals #1-33 was drilled in 1981, reached the bottom of the basalt flows at a depth of 5000 feet, then continued through sediments with numerous occurrences of coal, and some rhyolite tuff layers, to a depth of 16,199 feet.

We also note that at the southern edge of the *Columbia Basin Basalts* in north-central Oregon, the lava flows lie on top of the sedimentary John Day Formation; the John Day sediments contain abundant plant and land animal fossils.

It is apparent that the flows of the Columbia Basin did not follow one another in rapid succession, but that each flow was followed by a span of time prior to the next flow. In many cases, the span of time was sufficient for the formation of soil and the growth of plants, including mature trees in the Vantage Sandstone, and sufficient plant material for the formation of coal in many of the interflow zones. Additionally, the lowermost lava flows lie on top of rocks that include fossil-bearing sediments. The sequence observed is undeniable.

Conclusion

In Chapter 1 we discussed the nature of science, emphasizing that our scientific explanations are based on observations, and that our scientific explanations are tested, and validated or refuted, by further observations. So let's take a look at two suggested scientific ideas that have been promoted in the publications that promote the view that Earth is a recent creation: 1) the Earth is not more than a few thousand years old, and 2) nearly all of the fossil-bearing rock layers of Earth's crust were deposited during a universal flood only a few thousand years ago. Let's evaluate those scientific explanations in the light of our observations of the *Columbia Basin Basalts*, as reported and described in this chapter and the references cited.

First let us consider the age of these deposits. Each of the 100 or so layers of basalt was extruded, spread over a large area in a short span of time, cooled and solidified. The deposit of a layer of sediment and soil following many or most of the lava flows implies an extended period of time for the deposit of sediment derived from the area surrounding the deposit of lava, and for the formation of soil. The presence of coal in many of those interflow sediments implies the growth and accumulation of plants during the span of time between flows. In the case of the Vantage Sandstone, the presence of large petrified tree trunks and stumps implies at least decades of time between successive lava flows, and the thickness of the sandstone layer implies a longer time. What do you think?

We should add that temperature measurements were taken in the wells that were drilled into and through the lava flows, and the temperature increase with depth is near the average for crustal sedimentary rocks. That means that the lava flows have been cooled completely; there is no residual heat remaining underground from all that volcanic activity, by contrast with, for example, the heat that produces the geysers and hot springs in Yellowstone National Park.

We also note that the entire lava flow sequence lies on top of sediments of the John Day Formation, and other sediments encountered in the wells drilled through the basalt. Those sediments were apparently deposited by normal Earth processes of weathering, erosion, and transport. Those sediments, moreover, contain evidence of plant growth in the form of coalified plant fragments and layers of coal.

Along with the conviction that sedimentation and molten lava extrusion in the past would have occurred in pretty much the same way that these processes occur at present, all the evidence we have considered so far suggests that these deposits are much more than several thousand years old. We reach this conclusion without considering evidence from quantitatively measuring ages of rocks with radioactivity. Measurements of age, using radioactive isotopes, have been done, and the results have been published in the scientific literature. Those measurements confirm that most of the lava flows occurred between 17.5 million and 11 million years ago.

Second, let us consider the suggestion that nearly all of the fossil-bearing rock layers in the world were deposited during a single, worldwide flood as described in Genesis 8-9. We note that nearly all of each of the 100 or so lava flows of the *Columbia Basin Basalts* cooled and solidified under air. Those areas where the lava flows encountered water, as indicated by the formation of pillow lava, are a very small fraction of the total area covered by each lava flow. We note that there is an abundance of plant material in the interflow zones, as indicated by the occurrence of coal, and by the presence of spores and pollen in the coal samples from Rattlesnake Hills No. 1 Well. There is even one example of a mammal animal whose body was encased in the base of one of the lava flows, as described above. There are also many examples of plant and animal fossils from the John Day Formation, which is known to underlie the lava flows of the *Columbia Basin Basalts* in north central Oregon.

Certainly, the fossils mentioned above were not deposited in a single worldwide flood. The sediments in which the fossils are found lie below and between numerous lava flows that were extruded under air, not under water. The evidence - the observations - decisively refutes the "flood geology" suggestion.

So, if truth be known, there is much evidence, aside from measurements of ages of rocks by radioactivity, that leads to the conclusion that the rocks that form the Earth's crust are much more than a few thousand years old, and that the ancient deposits of sediments and volcanic rocks occurred by natural processes very similar to those that lead to deposits of sediments and volcanic rocks in our present-day world.

References

[1] G.F. Beck, "Fossil-bearing basalts, more particularly the Yakima basalts of Central Washington," *Northwest Science* 9, No. 4 (1935), 4-7.

[2] Beck, "Remarkable west American fossil, the Blue Lake rhino," *Mineralogist* 5, No. 8 (1937), 7-8 and 20-21.

[3] W.M. Chappell, J.W. Durham and D.E. Savage, "Mold of a Rhinoceros in Basalt, Lower Grand Coulee, Washington," *Geological Society of America Bulletin*, 62 (1951), 907-18.

[4] Bates McKee, *Cascadia*. (New York: McGraw-Hill, 1972), 279.

[5] J.R. Raymond and D.D. Tillson, *Evaluation of a Thick Basalt Sequence in South Central Washington*. BNL Report 776. (Richland, WA: Battelle Pacific Northwest Laboratories, 1968).

[6] Raymond, *Basalt Sequence*, 22.

[7] Raymond, *Basalt Sequence*, 70.

16 Radioactivity and Radioactive Decay

Beginning in the early 1900's, the radioactivity that is found naturally in rocks in the crust of the Earth has been used for measuring the ages of rocks, a procedure commonly called "radiometric dating." The results of those measurements indicate that Earth as a planet is about 4.65 billion years old, that rocks containing hard skeletal parts of fossil animals range from recent to 570 million years old, and that fossil algae and bacteria are found in rocks up to at least 3 billion years old.

Those who believe that Earth is a recent creation, not more than several thousand years old, and who also are trying to find scientific support for that viewpoint, reject the results of radiometric dating for any object older than those several thousand years. Many attempts to discredit the results of radiometric dating have been published, and an appreciable number of faulty claims are found in those publications. We should examine and evaluate some of those claims, since they have been distributed widely in the Christian community.

We will concern ourselves with those evaluations in the next several chapters. Our first task is to gain some basic understanding of the phenomenon known as "radioactivity."

Background

Within a few years after radioactivity was discovered by Henri Becquerel in 1896, the process known as "radioactive decay" was found to be time-dependent, that is, the process takes place at a regular, measurable rate. Since some rocks contain radioactive materials, the suggestion was made in 1905 that perhaps radioactivity could provide a way to measure the ages of rocks. That suggestion was soon put into practice; the publication of a professional paper by

Arthur Holmes in 1911 reported measurements of the ages of several rock samples, using the amounts of uranium and lead in those samples as the basis for age calculations. Some of the rocks studied by Holmes were found to be more than one billion years old.[1]

Determining ages of rocks and other materials by radiometric techniques has proven to be very important for learning about the history of our world and its inhabitants. The methods have been refined since 1911, and the results of the measurements have become widely accepted as valid.

First we must digress a bit to provide some basic information about radioactivity and the procedures that are used to measure ages of things. I want to tell you, the reader, enough so that you can understand that the results you have heard about and read about are not obtained by some magical process, nor are they the product of wishful thinking dredged up out of someone's imagination, nor are the numbers pulled down from skyhooks too lofty for ordinary people to understand, but the results are the product of the diligent and thoughtful work of conscientious people. We may have to define and use some new or unfamiliar words and terms, but I don't intend to "snow" you with a lot of technical stuff or fancy mathematics.

Atoms and atomic structure

The material stuff in the Earth, planets, stars, and galaxies consists of elements, either individually or in some combination. Gold and aluminum are individual elements, water is a combination of the elements hydrogen and oxygen, table salt is a combination of the elements sodium and chlorine, etc.

We can divide stuff into very small amounts or particles. When you dissolve some table salt in water, for example, the particles of salt are too small to see, even with an optical microscope, but you know the salt is still there by the taste. The smallest particle of an element that we can get that is still that element is called an "atom" of that element. Atoms in turn are made up of still smaller particles: protons and neutrons are found in the nucleus of the atom, and electrons are in motion surrounding that nucleus.

As noted above, each element is commonly identified by its name, such as "oxygen" or "iron" or "uranium," and those names may be referred to by a chemical symbol, such as "O" for oxygen, "Fe" for iron (from the Latin name, "ferrum"), "U" for uranium. In addition, each element is known by a unique number called the "atomic number." The atomic number of an element is the number of protons in the nucleus of an atom of that element. Thus, the element with atomic number 8 is oxygen, atomic number 26 is iron, atomic number 92 is uranium, etc.

Isotopes

It was discovered as early as 1907 that the atoms of any particular element are not all identical in mass. Those atoms of differing mass have come to be known as "isotopes" of that element. The language used to distinguish isotopes from each other makes use of a number called the "atomic mass number" accompanying the name or symbol of the element; for example, Co-60 and Co-59 are isotopes of cobalt, C-12, C-13, and C-14 are isotopes of carbon, etc.

So why must we talk about isotopes in connection with radioactivity? Let's consider the element iron as an example.

All isotopes of iron are alike in some ways: each atom of every isotope of iron has 26 protons in the nucleus, and each atom, when not part of a chemical compound, has 26 electrons surrounding the nucleus. All isotopes of iron behave alike in chemical reactions, as, for example in the rusting of iron in the presence of oxygen, forming iron oxide.

Continuing to use the element iron as our example: different isotopes of iron differ in mass, and these different isotopes can be separated from each other with an instrument called a "mass spectrometer." Different isotopes of iron differ in the number of neutrons in the nucleus, which accounts for the difference in mass.

More importantly for this discussion, some of the isotopes of iron are stable, that is, not radioactive, and some of the isotopes of iron are radioactive.

The phenomenon called "radioactivity" consists of processes that are taking place in the nucleus of the atom of a specific isotope of the element being considered. Different isotopes of each element have different characteristics with regard to those processes of radioactivity, so we must talk about each isotope separately in order to talk about its radioactivity.

Radioactivity and radioactive decay

The process called "radioactive decay" is the spontaneous transformation of an atom of a radioactive isotope into an atom of a different isotope, most commonly changing into an isotope of a different element. The radioactive isotope undergoing decay is called the "parent" and the resulting isotope is called the "daughter." ("Parent" is without gender, and there are no sons.) If the daughter is not radioactive, the process stops there; if the daughter is radioactive, it in turn becomes a parent and undergoes decay into another isotope, etc. until the eventual daughter is not radioactive.

Radioactive decay is a "time-dependent" process, that is, it occurs at a regular rate. The rate of decay of a radioactive isotope may be expressed in either of two ways: 1) as the probability that an atom of that isotope will undergo decay in a given amount of time, or, 2) as the half-life of the radioactive isotope, that is, the amount of time elapsed while half of the atoms of that isotope in a particular sample undergo decay.

The graph in Figure 5 describes the relationship of half-life and radioactive decay in two ways: 1) by the line graph, and 2) by the information associated with selected points on the graph. (Look at the graph in connection with each of the following sentences.)

If we begin with a sample containing 10,000 atoms of a radioactive isotope, as in this example, 5000 of them will undergo decay into atoms of the daughter isotope during one half-life, leaving 5000 unchanged (half of the initial number). During the next half-life, half of the remaining atoms of that radioactive isotope, 2500 of them, will

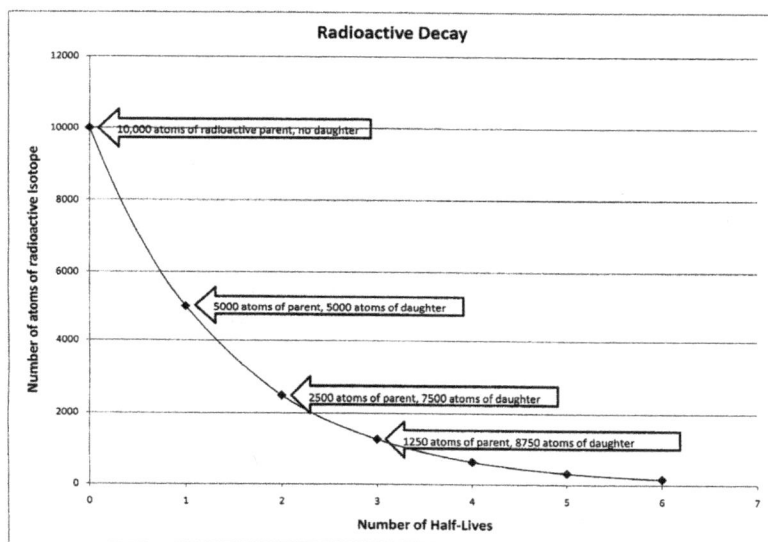

Figure 5. Radioactive decay.

decay into atoms of the daughter isotope, leaving 2500 unchanged. So, at the end of two half-lives, we still have half of half, or one-fourth of the initial number of atoms of the radioactive isotope in the sample. During the third half-life, half of those, 1250 of them, will undergo decay, leaving 1250 unchanged (half of one-fourth, or one-eighth of the initial number), etc. Note that the sum of the number of atoms of parent and the number of atoms of daughter at any point along the line equals the initial number of atoms of the radioactive parent, assuming that the daughter isotope is not radioactive.

The probability of radioactive decay and the half-life are obviously related; the higher the probability of decay, the shorter the half-life; the lower the probability of decay, the longer the half-life.

Isotopes commonly used in measuring ages of things

There are more than two thousand known radioactive isotopes of the slightly more than one hundred known elements. Their half-lives range from a small fraction of a second to several billion years. Some of the isotopes commonly used for measuring ages of rocks are: uranium-238, which decays with a half-life of 4.5 billion years, going through a series of steps into lead-206; uranium-235, which decays with a half-life of 700 million years, going through a series of steps into lead-207; thorium-232, which decays with a half-life of 14.1 billion years, going through a series of steps into lead-208; potassium-40, which decays with a half-life of 1.28 billion years, changing into calcium-40 (89% of decay events) or argon 40 (11% of decay events); and, for more recent materials that were once part of a living organism, carbon-14, which decays with a half-life of 5730 years, changing into nitrogen-14. In the next chapter, we will consider a couple of those parent-daughter isotope pairs in connection with measuring ages of things.

Isotopes in nature

Let us return to a discussion of isotopes of iron, as a typical example of what we find in nature, and what we have learned about isotopes in the laboratory. The iron that we find in the crust of the Earth consists of four stable isotopes: 5.8% is Fe-54, 91.8% is Fe-56, 2.1% is Fe-57, and 0.3% is Fe-58; all of these isotopes are stable, that is, non-radioactive.[2] In addition, 21 radioactive isotopes of iron have been identified; the atomic masses, half-lives, and other properties of each of these radioactive isotopes have been determined and published.[3]

Like iron, most of the elements as we find them on Earth consist of a mixture of a few or several stable isotopes. Elements with atomic number 1 through 83 have at least one stable, that is, non-radioactive, isotope (with two exceptions: technetium, #43, and promethium, #61). All elements have some radioactive isotopes; hydrogen has the fewest known radioactive isotopes with one, and many elements are known to have more than 20 radioactive isotopes. All elements with atomic number of 84 or higher, as well as technetium and promethium, have no stable isotopes, but only radioactive isotopes.

There are more details known about radioactive isotopes and their decay, but we won't take time or space to review those details here. Some of them are described in greater detail in Appendix B. We will run into a couple more details in some of the chapters to follow, and we will talk about them when we get there.

As noted earlier, radioactive decay is a "time-dependent" process, that is, it takes place at a regular rate. Therefore, like the rotation of the Earth, or the swing of a pendulum, or the flow of sand through the narrow constriction of an hourglass, the process can be used to measure the passage of time.

References

[1] Arthur Holmes, "The association of lead with uranium in rock-minerals, and its application to the measurement of geological time," *Proceedings of the Royal Society of London*, series A, vol. 85 (1911), 248-56.

[2] F. William Walker, George J. Kirouac, and Francis M. Rourke, *Chart of the Nuclides*. (San Jose, CA: General Electric Company, 1977).

[3] David R. Lide (Editor-in-chief), "Table of the Isotopes" in *CRC Handbook of Chemistry and Physics*. (Boca Raton, London, New York, Washington: CRC Press, 2003) Section 11, 50-197.

17 Measuring Ages of Rocks

Following the publication of that "landmark" paper by Holmes in 1911, reporting the ages of several rock samples on the basis of the amounts of uranium and lead in the samples, additional procedures have been developed and refined for measuring the ages of rocks by the radioactivity that is found naturally in rocks of the Earth's crust. The basic principle is straightforward, and will be presented here in simple prose. However, the details of applying that principle to measuring ages of actual rocks are sometimes a bit more complicated. A brief description will be provided in this chapter, and some additional information is found in Appendix C.

Basic Principle

Figure 6. Hourglass

The use of radioactivity to measure the age of a rock sample has often been said to be like measuring the passage of time with an hourglass. While true, the similarities can be observed only if we make the comparison with an hourglass that has sand running from top to bottom, and has some sand accumulated in the bottom while some sand remains in the top, like the hourglass pictured in Figure 6, and only if we ask the appropriate question.

So let us suppose that we come into a room in which an hourglass like the one pictured here is standing on the mantel. The room contains no other clocks or timepieces, and there is no other person in the room.

We ask, "I wonder how long this hourglass has been running?" (That is, how long has it been running since it was turned over, with all the sand initially in the top?)

Could we find out the answer to our question with the information we have, or information we can get from the hourglass itself, without asking the person who turned the hourglass over? With a little thought, and a little effort, yes, we can find out. (We may want to turn the hourglass on its side so the sand will stop running through the constriction, to preserve the information at the time when we asked the question.)

So, how would we find the answer to our question? We need three pieces of information:

1. The amount of sand remaining in the top of the hourglass at present.

2. The amount of sand in the top when it began running.

3. The rate at which sand flows from top to bottom.

With those three pieces of information, we can calculate how long the hourglass has been running.

So what we need to do is this:

1. We measure the amount of sand remaining in the top of the hourglass at the present time. (We can choose which units to use in our measuring: count individual grains of sand, or measure the amount in ounces, or grams, or ... whatever we choose.)

2. We measure the amount of sand that is in the bottom section of the hourglass at the present time.

3. We measure the rate at which the sand runs through the constriction between top and bottom when the hourglass is in operation.

Three measurements, which we are able to do. If we want to do those measurements precisely, we may have to use some tools to take the hourglass apart, and perhaps use some precision measuring devices, such as a scale or balance, to determine the amount of sand in the top and in the bottom at the present time. We will also need to use a timekeeping device, such as a watch or clock, to measure the rate at which sand flows from top to bottom, in order to be able to translate the amount of sand that has flowed through the constriction into minutes, or hours.

Then we need to do two calculations:

1. We can't measure the amount of sand that was in the top of the hourglass when it began running; we weren't there at the time. But we can find out how much was there. We add the amount of sand that remains in the top of the hourglass at the present time (measured in step 1) to the amount of sand that has run through the constriction into the bottom (measured in step 2); the sum is the amount that was in the top initially.

2. Now we can do the final calculation. The amount of time that has passed since the hourglass began running equals the amount of sand that has flowed through the constriction, that is, the difference between the amount of sand that was in the top initially and the amount that is in the top at present, divided by the rate at which the sand flows through the constriction. (We recognize that the difference in the amount of sand in the top from initial to present is just the amount of sand in the bottom that we measured in measurement #2 above. However, I stated it as a difference since that is the approach that we find useful in measuring the passage of time using radioactivity.)

Three measurements, which we are able to do at the present time, that is, at the time of our observation and measurement; we didn't have to be there when the hourglass began running.

Two calculations, not too difficult.

Similarly, to determine the age of a rock sample using a particular radioactive isotope, we need to know three things:

1. The amount of the radioactive isotope that is in the rock sample at the present time (analogous to the amount of sand in the top of our hourglass analogy at the present time).

2. The amount of the radioactive isotope that was in the rock sample when the rock was formed (analogous to the amount of sand in the top of our hourglass analogy when the hourglass began running).

3. The rate at which the atoms of that radioactive isotope change into atoms of the daughter isotope, or the half-life of the radioactive isotope (analogous to the rate at which sand flows through the constriction in our hourglass analogy).

With those three pieces of information we can calculate the age of the rock, that is, how long the radioactive parent isotope in the rock sample has been decaying into its daughter isotope since the rock was formed. (We don't have to stop the process of radioactive decay in the rock sample while we do some measurements, as we had to do with our hourglass analogy, because the half-lives of the isotopes we use are very long, and change takes place very slowly.)

So how do we go about getting that information?

1. Finding out how much of the radioactive isotope is in the rock sample at the present time is easy; we measure it. ("Easy" is, I suppose, a relative term. We need to have the appropriate tools available, and the measuring takes skill, of course, and skills must be learned, etc. etc. But it is straightforward.)

2. Finding out how much of the radioactive isotope was in the rock sample when the rock was formed is a somewhat greater challenge - we weren't there at the time. (We weren't there when our hourglass began running, either.) However, we know that the atoms of the radioactive isotope that have decayed during the history of the rock have changed into atoms of the daughter isotope (like each grain of sand in the bottom of our hourglass has flowed from the top of the hourglass through the constriction and into the bottom during the running of the hourglass). So what we must do is determine the number of atoms of daughter isotope that have been produced in the

rock sample from the decay of the radioactive parent isotope. We know that the number of atoms of radioactive isotope in the rock sample when it was formed is the sum of those in the rock sample at the present time and the number of those that have decayed into atoms of daughter isotope in the rock sample since the rock was formed.

3. Many measurements have been made to determine the half-lives of the radioactive isotopes involved, so we may be confident that those numbers are well known. If we haven't gotten that information by our own measurement, we can look it up in a book.

That's the basic theme. It's totally reasonable, and straightforward, as in the hourglass analogy.

We should note that if we were to prepare a graph showing the change in the amount of sand in the top of the hourglass over time, we would get a straight line, whereas in the graph for radioactive decay that was shown in the previous chapter we get a shape known as "exponential" decay. Either one can be used for measuring the passage of time, but the mathematics is slightly different in the two cases.

The basic principle is simple and straightforward. There are some variations on the measurements and calculations that have to be made with different parent-daughter isotope pairs, but the principle remains the same.

Conditions necessary for valid age measurement - sample selection

In our hourglass analogy above, it is obvious that the sand inside the hourglass cannot get out, and no sand from elsewhere can get into the sealed glass container. If any sand were to get lost from the hourglass, or added to it, during the time that it has been running, the result of our calculations would be in error. Similar conditions apply for valid age measurements of rocks with radioactive isotopes.

1. There must have been no radioactive parent isotope added to,

or lost from, the rock sample during its history, except by the processes of radioactive decay.

2. Also, for most of the specific procedures, there must have been no daughter isotope added to, or lost from, the sample during its history, except by the process of radioactive decay of the parent. (Carbon-14 is an exception; the amount of daughter isotope is not considered in that method, as will be explained later.)

In other words, we must have a closed system in order for the sample to be suitable for measuring its age with radioactivity. Without doubt, proper sample selection is the most important factor in judging whether those conditions are met, or nearly met, for a particular rock.

Many experiments have been carried out to investigate what kinds of changes might occur in different kinds of rocks under a wide range of conditions, and what sorts of events might result in the migration of various isotopes into, out of, or within a rock sample. The results of those studies indicate that the necessary conditions <u>are</u> met, or very nearly so, for minerals in crystalline igneous rocks, rocks that have been formed by crystallization from molten magma under such conditions that the rock consists of distinct crystals of its constituent minerals.

The basis for that confidence is found in the nature of mineral crystals and the process by which crystalline igneous rocks are formed. When molten magma cools and solidifies slowly enough, the elements present in the magma form chemical compounds, and these compounds solidify into crystals of various minerals. The atoms in these mineral crystals are arranged in a regular geometric pattern. You have probably seen the expression of those geometric patterns in the shapes of mineral crystals in museums, or in school collections. The elements found in the molten magma in the greatest abundance will form mineral crystals that make up most of the rock; these are called "major" minerals. Granite, for example, consists mostly of crystals of quartz, feldspar, and mica. Elements found in the molten magma in smaller quantities may form mineral crystals in lesser abundance than the major minerals present; these are called "accessory" or "minor" minerals. Elements found in the molten magma in still smaller quantities are incorporated into other mineral crystals as impurities, with the atoms of those elements occupying sites in the crystal structure as "replacements" for atoms of elements in the major or minor minerals that are nearly the same size as the replacement atoms.

After the resulting rock has solidified and cooled to the temperature at or near the Earth's surface, the atoms in the mineral structure of the rock are fixed in place. When an atom of a radioactive isotope decays, the atom of its daughter isotope is trapped in the same site in the crystal that had been occupied by the radioactive parent atom. Since crystalline igneous rocks are impervious to penetration by groundwater, and since they are weathered slowly from exposed surfaces, the mineral crystals in the unweathered rock almost certainly have neither lost nor gained any atoms by interaction with their surroundings throughout their history.

We know that there are some types of rocks for which the conditions for reliable age measurement by radioactive methods are very likely not met. For example, sedimentary rocks such as shale, sandstone, and limestone are often somewhat porous, allowing ground water to flow through them; groundwater may dissolve minerals away from the components of the rock, or add minerals that were in solution in the groundwater. While some experiments have been done to investigate the possibility of radioactive age measurements on certain unusual minerals in sedimentary deposits, sedimentary rocks are not usually considered to be sources for samples whose age can be measured reliably with radioactive isotopes. We will consider additional concerns with sample selection in our discussion in later chapters.

Procedures

So that you, the reader, will be able to gain some understanding of the reliability of the methods, we will consider the main outlines of the procedures for two of the isotope pairs in ordinary prose in this chapter. Additional details of the procedures will be presented in more technical language in Appendix C. If you wish to do so, you can learn about procedures that have been developed, or are under development, for a larger number of isotope pairs by consulting books on age measurement with radioactive isotopes.[1]

Potassium-40/Argon-40

Potassium is a component of many rock-forming minerals found in Earth's crust. A small fraction, 0.012% (12 out of every 100,000 atoms), of naturally occurring potassium is K-40, which is radioactive. K-40 has a half-life of 1.28 billion years.

Potassium-40 decays in either of two ways: 1) 89% of the K-40 atoms decay into atoms of calcium-40, and 2) 11% of the K-40 atoms decay into argon-40 atoms. The decays resulting in Ca-40 are not useful for measuring ages because there is a large amount of Ca-40 in the same materials that contain potassium, and it would be very difficult to detect the addition of a few atoms of Ca-40 from K-40 decay. The decays resulting in Ar-40, however, enable us to measure the ages of properly chosen samples.

Argon is a "noble" element, that is, it does not form chemical compounds with any other elements except under very unusual, laboratory-produced conditions. Consequently, argon is excluded from the minerals in rocks formed from molten magma, if the magma cools and crystallizes under conditions that result in distinct mineral crystals. There is no space within the rock-forming mineral crystals for argon atoms. In crystalline igneous rocks, such as granite and gabbro, then, there is no argon present in the mineral crystals at the time the rock cooled and solidified. The situation is like that of an hourglass that began running with no sand present in the bottom of the glass.

There is some potassium present in several types of mineral crystals within many rocks. During the span of history since a crystalline igneous rock was formed, some of the K-40 atoms that were initially in a sample of that rock have undergone radioactive decay. Of those that have undergone radioactive decay, 11% will have formed atoms of Ar-40, and those argon atoms are trapped in the mineral crystal at the same locations that had been occupied by the K-40 parent atoms.

With that understanding of the formation of mineral crystals and the behavior of argon, we have a way of finding out the amount of potassium-40 that was present in the rock when it was formed. We make two measurements: 1) we measure the amount of K-40 that is present in the rock sample at present, and 2) we measure the amount of Ar-40 that is present in the rock sample at present. Taking into account the fact that the Ar-40 in the rock at present is the product of 11% of the decays of K-40 atoms, we can determine the total number of atoms of K-40 that have decayed during the history of the rock. The number of atoms of K-40 that were present in the sample when the rock was formed is the sum of those present in the rock at present and those that have undergone radioactive decay during the history of the rock.

We can now calculate the age of the rock. We can do that graphically by taking two quantities, 1) the number of atoms of K-40 in the rock sample when the rock was formed, and 2) the number of K-40 atoms that remain in the rock sample at the present time, and fitting them to a graph like that presented in Chapter 16, with one half-life of K-40 equal to 1.28 billion years. For example, if half of the atoms of K-40 that were in the rock when it was formed have undergone radioactive decay, one half-life of K-40 has elapsed since the rock was formed, and the rock is 1.28 billion years old. Or, alternatively, we can use the mathematical equation that describes the decay pattern that is represented in such a graph, and calculate the time elapsed during the history of the rock from those two quantities.

Many results of such measurements and calculations have been reported in the scientific literature.

The matter of sample selection for the K-40/Ar-40 method is important, since we know that some samples of rocks formed by recent volcanic activity have some Ar-40 trapped in glassy (that is, non-crystalline) portions of the rock, apparently because some Ar-40 had been dissolved in the molten magma from which the rock was formed. This subject will be considered in detail in Chapter 24.

Uranium-238/Lead-206

The landmark paper by Arthur Holmes in 1911 calculated the ages of the rock samples he had collected on the basis of measurements of the total amount of uranium and the total amount of lead in the samples. The results reported by Holmes are pretty good, because nearly all of the uranium in Earth's crust consists of U-238, and most of the lead in Holmes' samples consisted of Pb-206. Today the determination of ages makes use of specific isotopes of various elements, rather than using the total amounts of those elements in the rock samples.

Uranium-238 undergoes radioactive decay through a series of steps, ultimately ending in non-radioactive lead-206. The intermediate steps are radioactive isotopes with shorter half-lives than that of U-238, so we need only consider the two endpoints of the series, U-238 and Pb-206. The half-life of U-238 is 4.5 billion years.

In the case of potassium-40 and argon-40, we can find samples for which we may be confident that there was none of the daughter isotope in the rock sample when it was formed. However, we cannot have that same confidence for U-238 decaying to Pb-206. Some lead was incorporated into many, perhaps most, of the rocks in the crust of the Earth when they were formed. So, if we want to use U-238 and Pb-206 for measuring the age of a rock, we will have to find some way of finding out how much Pb-206 was in the sample when the rock was formed, so that we can determine how much of the Pb-206 in the sample at the present time is the result of the radioactive decay of U-238. And, yes, as you might have guessed, there is such a way, or we wouldn't be using that isotope pair for age measurements.

The lead in the crust of the Earth consists of four non-radioactive isotopes of lead; on average, the lead consists of 1.42% Pb-204, 24.1% Pb-206, 22.1% Pb-207, and 52.4% Pb-208. The amounts of Pb-206, -207, and -208 in the crust of the Earth are increasing over time; Pb-206 from the radioactive decay of U-238, Pb-207 from the radioactive decay of U-235, and Pb-208 from the radioactive decay of thorium-232. However, there is no radioactive isotope in Earth's crust that decays into Pb-204, so the amount of Pb-204 in the crust of the Earth, and in any rock sample, remains constant over time. So, instead of using the <u>amount</u> of Pb-206 in a rock sample to measure the age of a rock containing U-238, we use the <u>ratio</u> of Pb-206 to Pb-204. Since the denominator of that ratio, the amount of Pb-204, remains constant over time, an increase in the ratio of Pb-206 to Pb-204 is due to the addition of Pb-206 from the radioactive decay of U-238.

In the most straight-forward case, we choose two minerals from a crystalline igneous rock sample: 1) a mineral that contains little or no uranium, and 2) a mineral that contains some uranium. We measure the ratio of Pb-206 to Pb-204 in both minerals. The ratio of Pb-206 to Pb-204 has not changed in mineral #1 since the rock was formed, because no uranium is present in that mineral to produce additional Pb-206. In mineral #2, however, the radioactive decay of U-238 has added some Pb-206, increasing the ratio of Pb-206 to Pb-204. The amount by which the ratio of Pb-206 to Pb-204 has changed in mineral #2 provides the information we need to determine how much U-238 has decayed during the history of the rock.

The amount of U-238 that was in the rock sample when the rock was formed is the sum of the amount that is in that sample at the present time and the amount of U-238 that has decayed during the history of the rock. Now we can use the graphical method suggested earlier, or do the calculation with the mathematical equation, to determine the age of the rock.

Here are a few of the results achieved by measuring the ages of rocks with radioactivity, in addition to those reported in the first paragraph of chapter 16: 1) the oldest rocks that have been found on Earth were formed at least 3.8 billion years ago; 2) rocks returned from the moon by Apollo space travelers range in age up to 4.15 billion years; 3) dinosaur fossils are found in rocks ranging from 245 million years old to 66 million years old, but have not been found in rocks younger than that.

The perspective of publications that promote the view that Earth is a recent creation

Needless to say, advocates of the view that Earth is a recent creation, and who also are trying to support that view with the results of the scientific investigation of God's world, do not find such results acceptable, and arguments against the validity of those results are found throughout the publications that promote that view.

Let us consider here one such argument, namely, the claim that all radiometric age measurement is based on faulty assumptions. The book *Scientific Creationism*, promoting the view that Earth is a recent creation, contains a lengthy discussion of radiometric age measurements. Some of the specific methods for age measurement are listed, and the following claim is made (italics in the original):

"For these or other methods of geochronometry, one should note carefully that the following assumptions must be made:
1. *The system must have been a closed system.*
"That is, it cannot have been altered by factors extraneous to the dating process; nothing inside the system could have been removed, and nothing outside the system added to it.
2. *The system must initially have contained none of its daughter component.*
"If any of the daughter component were present initially, the initial amount must be corrected in order to get a meaningful calculation.
3. *The process rate must always have been the same.*"[2]

"Assumption" #1. from *Scientific Creationism* is a necessary condition for valid radiometric age measurement; we have discussed it earlier. We will take up "Assumption" #3 in Chapter 19.

"Assumption" #2, the condition that the rock sample "initially contained none of its daughter component," is required for the K-40/Ar-40 method, as we have noted above. However, there are many radiometric age methods for which the condition of "initially contained none of its daughter component" is neither assumed nor required, as, for example, in the U-238/Pb-206 method described earlier. It is true that corrections must be made in some of the methods for determining the initial amounts of the daughter isotope; as we have seen in the example of the U-238/Pb-206 method, there are reliable procedures for making such corrections.

The discussion in *Scientific Creationism* continues with judgments regarding the validity of those three "assumptions," and we will pay particular attention to the published judgment regarding "assumption" #2., which is stated as follows:

"It is impossible to ever know the initial components of a system formed in prehistoric times.

"Obviously no one was present when such a system was first formed."[3]

If truth be known, as we have seen and shall see, reliable ways have been devised to find out what the initial isotopic composition was when the rock was formed. Like any investigation of historical events, these methods are based on evidence available to us at the present time, coupled with sound logic. We didn't have to be there when the rock was formed.

We noted in our hourglass analogy that we didn't have to be there when the hourglass was turned over in order to find out how long the hourglass had been running. There are many other analogies available from any study of history in which the historian does not have eyewitness testimony or written records from the time of the event. In just one such example, consider the challenge faced by an experienced hunter/tracker, who studies animal tracks to learn some things about his/her quarry; the hunter wasn't there when the animal made the tracks, but he or she can still learn a great deal about the animal and its behavior.

There are some Christians who find the results of radiometric dating unacceptable, believing that Scripture teaches otherwise. They are entitled to that judgment, of course. We are not better Christians if we accept the validity of the results of radiometric age measurement, nor are we worse Christians if we do not, just as we are not better Christians if we think that the Earth rotates and revolves around the sun, nor are we worse Christians if we think the Earth is stationary in space while sun, moon and stars revolve around the Earth. Romans 10:9 assures us that "if you confess with your mouth 'Jesus is Lord,' and believe in your heart that God raised him from the dead, you will be saved." The ages of rocks and other matters of our scientific understanding of the structure and behavior of God's physical world are not vital to our Christian faith, though they may be of more or less interest to us for various reasons.

References

[1] Alan P. Dickin, *Radiogenic Isotope Geology*. (Cambridge, UK: Cambridge University Press, 2005).

[2] Henry M. Morris, Ed., *Scientific Creationism*, (San Diego, Creation-Life Publishers, 1974), 238.

[3] Morris, *Creationism*, 239.

18 Quotations and Misquotations: Uranium-Lead Methods

As we noted in Chapter 9, it is common practice to use quotations from other authors in articles in journals, in discussions in books, in public lectures, etc. The commonly accepted standards for such quotations were listed in Chapter 9.

The discussion of radiometric age measurement of rocks in *Scientific Creationism*[1] lists several claimed flaws in the procedures used to determine such ages. One of the requirements for such procedures to be valid is that none of the parent radioactive isotope or the daughter isotope of interest be lost from, or added to, the sample during the history of the rock, except by the processes of radioactive decay. A sample which meets that requirement is called a "closed system." The section of *Scientific Creationism* that we consider here is the claim that this requirement is not generally met by samples used for measuring ages of rocks. The chapter in *Scientific Creationism* that we consider is entitled "Old or Young?" and the major section in question is entitled "Radiometric Dating" with the sub-heading "1. The Uranium Methods." The specific claim is expressed as follows:

"(a) Uranium minerals always exist in open systems, not closed.

"Uranium is easily leachable by groundwater, for example. The intermediate element, radon gas, can easily move in or out of a uranium system. There are, in fact, various ways by which the components of this type of system can enter or leave it. One of the chief authorities on radioactive dating, Henry Faul, said:

"'Uranium and lead both migrate (in shales) in geologic time, and detailed analyses have shown that useful ages cannot be obtained with them. Similar difficulties prevail in attempts to date pitchblende veins. Here again much chemical activity is known to take place and widely diverging ages can be measured on samples from the same spot.'"[2]

The quotation from Henry Faul is from the book *Ages of Rocks, Planets and Stars*. The passage in Faul's book from which the quotation is taken is part of a discussion of the kinds of samples that are useful for measuring ages of rocks by radioactivity, specifically those that are found in contact with sedimentary rocks, so that the ages of sedimentary rocks may be inferred from that information. After reviewing various requirements for reliability of results, Faul describes three types of igneous rock deposits that very nearly satisfy the requirements of being "closed systems," and are sometimes found in contact with sedimentary rocks, specifically, "layered volcanics," "bracketed intrusives," and "simple intrusives." He considers results for such samples to be valid for establishing ages of sedimentary formations in the "geologic column" that have been published in the professional geology literature.

Then Faul adds a section with the heading "Questionable reference points." This passage is about sample selection, and not about the reliability of the procedures used. The passage that is quoted in *Scientific Creationism* is taken from this section, which deals with several types of samples that have been recognized as unreliable. Such samples are not used as a basis for the ages of rocks reported in the current geological and geochemical literature, of course.

The quotation, as published in *Scientific Creationism*, appears to support the claim that age measurements using uranium and lead isotopes are not reliable. But let's take a somewhat more careful look at the quotation and compare it with the original passage from Faul's book.

The full passage in the original source is as follows:

"Uraniferous shale is another unreliable system. In several parts of the world are large shale deposits with fairly high uranium contents. Their stratigraphic position is accurately known, but these rocks are not closed systems. Uranium and lead both migrate in them in geologic time, and detailed analyses have shown that useful ages cannot be obtained from them. Similar difficulties prevail in attempts to date pitchblende veins.

Here again much chemical activity is known to take place and widely diverging ages can be measured on samples from the same spot."[3]

Note that the quotation as printed in *Scientific Creationism* (reprinted above) is only a part of the full passage in Faul, and that it is not a *verbatim* quotation. The original passage has been altered in more than one way, although it is enclosed in quotation marks in *Scientific Creationism* as if it were a *verbatim* copy of the passage in Faul. Note the following:

1. The first three sentences of the paragraph in the original passage from Faul are not included in the quotation in *Scientific Creationism*.

In many cases, the preceding sentences would not be essential to the meaning of a passage that is quoted from a different publication. In this case, however, it is critical because the meaning of the fourth sentence, which is quoted in *Scientific Creationism*, is dependent on those first three sentences.

In the original paragraph in Faul, the subject of the first sentence is "Uraniferous shale;" in the second sentence he refers to the same kinds of deposits as "shale deposits with fairly high uranium contents;" and in the third sentence he refers to the same kinds of deposits as "these rocks." Then follows the fourth sentence:

"Uranium and lead both migrate in them in geologic time, and detailed analyses have shown that useful ages cannot be obtained from them."

By way of contrast, the quotation in *Scientific Creationism* reads as follows:

"Uranium and lead both migrate (in shales) in geologic time, and detailed analyses have shown that useful ages cannot be obtained with them."

The pronoun "them," used twice in that fourth sentence in the original passage, clearly refers to the "uraniferous shale" that Faul has spoken of in the first three sentences. The omission of those first three sentences, and alterations in the text in the quotation in *Scientific Creationism* are likely to lead the reader to conclude that "them," used only once in the quotation, refers to "uranium and lead."

2. The text of the original passage has been altered in the quotation in *Scientific Creationism* in ways that may readily lead to a misunderstanding for all but the most careful readers.

"Uranium and lead" is certainly the subject of the verb "migrate" in both the fourth sentence of the original passage in Faul and in the quotation in *Scientific Creationism*. In the quotation in *Scientific Creationism*, however, the original "in them," referring to the uraniferous shales described in the preceding three sentences, has been replaced with the parenthetical expression "(in shales)," and the expression "from them," in the original passage in Faul has been changed to "with them." Those substitutions strengthen the potential impression that the pronoun "them" refers to uranium and lead, rather than to the uraniferous shale samples.

Clearly, Faul was speaking of the type of sample material in saying that "useful ages cannot be obtained from them." The out-of-context quotation in *Scientific Creationism*, with the changes in wording, gives the impression that Faul is criticizing the uranium-lead method, whereas, as is evident in the full passage in Faul's book, he is informing his readers of the need for care in sample selection.

If truth be known, the recognition that some types of rocks do not provide reliable results of age measurement by radioactivity, and understanding the reasons for such unreliability, allows investigators to choose rock samples that do meet reliability requirements. Knowing that the scientific investigators are aware of potential pitfalls actually justifies an increased confidence that the results reported in the secular scientific literature are correct.

You may choose to believe that the Earth is a recent creation, and that the creative acts recorded in Genesis Chapter 1 took place in six twenty-four-hour days, if you wish. But quotations from the professional literature that have been altered to change their original meaning, or that have been lifted from a context that actually argues against the claim being promoted, do not provide legitimate support for such a claim.

References

[1]Henry M. Morris, Ed., *Scientific Creationism.* (San Diego: Creation-Life Publishers, 1974).

[2]Morris, *Creationism*, 140.

[3]Henry Faul, *Ages of Rocks, Planets and Stars.* (New York: McGraw-Hill, 1966), 61.

19 Constancy of Radioactive Decay Rates

In order to be confident that the published results of measuring ages of rocks with radioactivity are reliable, we must have reason to be confident that the half-lives of the radioactive isotopes in use for the measurement are constant over long periods of time. After all, the published results include ages of some rocks in the billions of years.

Publications that promote the view that Earth is a recent creation reject the results of measurement of ages of rocks by radioactive isotopes, of course. These publications contain many claims that such measurements are unreliable. One such claim is that there is no assurance that radioactive decay rates are constant. This claim is presented and discussed in some detail in the book *Scientific Creationism*, as follows:

> "*No process rate is unchangeable.*
> "Every process in nature operates at a rate which is influenced by a number of different factors. If any of these factors change, the process rate changes."[1]
> "The uranium decay rates may well be variable.
> "Writers on this subject commonly stress the invariability of radioactive decay rates, but the fact is these rates, as well as all others are subject to change. Since they are controlled by atomic structure, they are not as easily affected as other processes, but factors which can influence atomic structures and processes can also influence radioactive decay rates."[2]

The assertion that radioactive decay rates might not be constant over time is also presented on the website www.icr.org/radioactive-decay-rates-not-stable/.

The paragraph in *Scientific Creationism* that immediately follows the quotation above suggests a couple of factors that might affect radioactive decay rates, specifically cosmic radiation and its production of neutrinos, and the possibility of bombardment of uranium-bearing rocks by neutrons. We will consider both of those in later chapters, but let us first consider the broader question: Is there evidence that gives us good reasons to think that radioactive decay rates are constant over billions of years in the Earth environment in which we and the rocks exist?

The quotation from *Scientific Creationism* above mentions that radioactive decay rates are controlled by atomic structures, and therefore are not easily affected by external factors. More specifically, radioactive decay rates are controlled by the structure of the <u>nucleus</u> of the atom, which is shielded from nearly all external influences by the electrons that surround the nucleus.

A number of experiments have been performed to examine the question of the constancy of radioactive decay rates. In some cases, certain changes in chemical structure or external pressure were found to have a small effect on radioactive decay rate, changing the rate by less than 1%. Such a small change would not alter the results of age measurements by radioactive decay to any significant extent. See Appendix D for further details and references.

Furthermore, for any change in radioactive decay rates resulting from changes in chemical structure to have any effect on age measurements, the change would have had to take place <u>during</u> the history of the rock. Since the radioactive isotopes in crystalline igneous rocks have been in the same chemical environment throughout the history of the rock, no change in decay rate due to a change in chemical structure would have taken place during that time.

Additional considerations

During the past couple of hundred years, measurements have been made of quantities now called "fundamental constants," or sometimes called "natural constants." These include such quantities as the gravitational constant, the amount (magnitude) of the electrical charge on an electron, the velocity of light in vacuum, and many more. These fundamental constants have been found to be interconnected with each other in a complex web of interrelationships. This complex web is sometimes called the "fine tuning" of the universe, because if any of these fundamental constants were even a little larger or smaller than it is, the universe would be a very different place. This feature of the universe has given rise to the "anthropic principle," which affirms that the existence of humans on Earth would be impossible in a universe with different values of the fundamental constants.

The rates of decay of radioactive isotopes are also interrelated with these fundamental constants. If the fundamental constants are truly constant over the history of the universe, the rates of decay of radioactive isotopes will also have remained constant over that span of time. There has been considerable interest in finding out whether or not the fundamental constants are truly constant, resulting in technical publications on the subject as early as 1937 and continuing to the present. Recent measurements have been able to determine an upper limit to the amount of change that has occurred in the value of the fundamental constants, and the upper limit of change over the lifetime of the universe is about one percent.[3] In other words, any change that has occurred in the fundamental constants is less than 1%, and may well be zero. We may be confident, then, that radioactive decay rates have not varied by more than 1% from their present measured values during the history of the universe.

You see, the idea that radioactive decay rates are constant is not just a bare-faced assumption. If truth be known, the conclusion that radioactive decay rates are constant, or nearly so, is based on careful study and extensive experimentation.

References

[1]Henry A. Morris, Ed., *Scientific Creationism*, (San Diego: Creation-Life

Publishers, 1974), 139.

[2]Morris, *Creationism*, 142.

[3]Donald Lindsay, "Evidence about constants being the same in the distant past," (includes bibliography) from the website www.don-lindsay-archive.org/creation/constant_evidence.html (2006).

20 A Sea of Neutrinos

In the discussion of uranium methods for measuring the ages of rocks in the book *Scientific Creationism*, in a subsection entitled "(b) The uranium decay rates may well be variable," the claim is published that the production of neutrinos by cosmic radiation could influence radioactive decay rates. That claim is elaborated as follows:

"Phenomena such as these [production of neutrinos] would be generated by such events as the reversal of the earth's magnetic field or supernova explosions in nearby stars. Since such phenomena are commonly accepted now as having occurred in the past, even by uniformitarian astronomers and geologists, there is a very real possibility that radioactive decay rates were much higher at various intervals in the past than they are at present. That this possibility is being considered seriously is evident from the following comment by Dr. Fred Jueneman, who is director of research for the Innovative Concepts Association."[1]

The quotation from Dr. Fred Jueneman is as follows:

"Being so close, the anisotropic neutrino flux of the super-explosion must have had the peculiar characteristic of resetting all our atomic clocks. This would knock our Carbon-14, Potassium-Argon, and Uranium-Lead dating measurements into a cocked hat! The age of prehistoric artifacts, the age of the earth, and that of the universe would be thrown into doubt."[2]

Do we have any evidence that might bear on that conjecture? Yes, a little. First, a few words about neutrinos. A neutrino is a particle with no electrical charge, and a rest mass of zero. Neutrinos are very difficult to detect because they interact very little with other matter. Techniques *have* been developed to detect them, however.

Scientific explanations developed in astrophysics indicate that processes going on in stars produce neutrinos, and a supernova explosion produces a very large number of neutrinos. Supernova explosions also produce gamma rays and visible light. For an observer on Earth, the differences in the arrival times of gamma rays, visible light, and neutrinos from a supernova event can be predicted on the basis of those explanations.

A fine opportunity to "test" those theories occurred in 1987 when a supernova explosion was observed (23-24 February) that had occurred in the Large Magellanic cloud, a star cluster visible from the southern hemisphere. The star cluster is 50 kiloparsecs (150,000 light years) from Earth.[3] The visible light from the explosion was observed by the Hubble telescope, and a burst of neutrinos was observed by two detectors on Earth (Japan and Ohio) at exactly the theoretically expected amount of time before the arrival of visible light from that supernova explosion.[4] Did you feel the neutrinos go through you? Calculations indicate that more than 10^{14} (1 followed by 14 zeroes, or 100 billion billion) neutrinos passed through each human adult on Earth from that event.

That "sea" of neutrinos had no noticeable effect on radioactive decay rates, nor any resetting of atomic clocks.

Chinese astronomers observed and recorded a supernova explosion in 1054 A.D. in which the visible light was as bright as the full moon, gradually fading to invisible to the naked eye over a period of about two months. The remnant of that explosion is known as the Crab Nebula. The distance from Earth to the Crab Nebula is 2 kiloparsecs (6,000 light years), much closer than the 1987 supernova in the Large Magellanic star cluster. No neutrino detectors existed at the time, so no measurements of neutrino flux are available. However, the distance to the Large Magellanic Cloud is 25 times the distance to the Crab Nebula, so the fraction of the total neutrino flux from the 1054 A.D. supernova that reached the Earth would have been about 625 (25-squared) times the fraction of the total flux that reached the Earth from the 1987 supernova. Carbon-14 measurements have been made on samples with known historical age back to the early Egyptian dynasties, and on tree rings back in time for more than 8000 years. The agreement between carbon-14 measurements and dates known from historical records or from tree ring data follows a smooth pattern throughout that time, and does not indicate any departure from the currently measured decay rate. There was no sudden spike or sudden change in the carbon-14 record at 1054 A.D. when the Crab Nebula supernova was observed. If truth be known, therefore, the evidence that we have argues against any effect on radioactive decay rates by the neutrinos from supernova explosions.

About Frederic B. Jueneman

We should add a word about the writings of Frederic B. Jueneman, a portion of which was quoted in *Scientific Creationism* as evidence that the possible alteration of radioactive decay rates by neutrinos is being "considered seriously."

Jueneman's columns were published in the monthly journal *Research & Development* (previously named *Industrial Research & Development*) over a span of more than two decades. The monthly column was entitled "Scientific Speculation." Selected columns were republished in two book-length collections, *limits of uncertainty*, subtitled "essays in scientific speculation," in 1975, and *Raptures of the Deep*, subtitled "Essays in Speculative Science," in 1995. The author's Preface to *limits of uncertainty* provides some insight into his perception of his own writing in those columns:

> "all were written to tickle the imagination"
>
> "These are by no means intended to be scholarly commentaries"
>
> "If there is any underlying philosophy in this collection, it is that I consider nothing is beyond question. Everything must be subject to scrutiny, right down to the most fundamental postulates."[5]

As I (CM) perceive it, the central theme running through Jueneman's columns is a questioning of the claims of science which have been adopted as "dogma" by at least some people, and such people then consider these claims to be beyond further questioning or investigation. However, Jueneman thinks (and I agree) that none of the claims of science is beyond questioning and further examination. The passage from *Scientific Creationism* which was quoted at the beginning of this chapter states that the possibility that radioactive decay rates would be affected by a sea of neutrinos was being given "serious consideration" by Jueneman and others. If truth be known, however, Jueneman refers to his own writing as "speculation." We should be careful not to confuse "speculation" with "serious consideration."

Nevertheless, we have given "serious consideration" to his speculation regarding the resetting of atomic clocks and the altering of radioactive decay rates by a "sea" of neutrinos, and the evidence that we have found indicates that massive amounts of neutrinos from supernova explosions have no perceptible effect on the rates of decay in radioactive isotopes on Earth.

References

[1] Henry M. Morris, Ed., *Scientific Creationism*. (San Diego, Creation-Life Publishers, 1974), 142-3.

[2] Frederic B. Jueneman, "Scientific Speculation," *Industrial Research & Development* (September 1972), 15.

[3] M. Mitchell Waldrop, "Supernova 1987A," *Science* 235, (6 March 1987), 1143; (13 March), 1322; (20 March), 1461; Vol. 236, (1 May), 522.

[4] John D. Fix, *Astronomy: Journey to the Cosmic Frontier*. (New York: McGraw-Hill, 2001), 462.

[5] Jueneman, *limits of uncertainty*. (New York: Cahners Publishing Company, 1975), Preface.

21 Carbon-14 Dating

The method known as radiocarbon dating was developed by Willard F. Libby in 1950, and has been in widespread use since then. The method is used to measure the ages of the remains of living organisms, and is especially useful for determining the ages of artifacts of the past history of human culture. The results indicate that humans have inhabited the Earth for at least 35,000 years, and probably longer. Those results are unacceptable to proponents of the view that Earth is a recent creation, whose publications have included several criticisms of the method. In this chapter and the next we will evaluate some of those criticisms.

We must begin with a review of the method. Carbon-14 is produced in the Earth's atmosphere by reactions of cosmic rays with the oxygen and nitrogen in Earth's upper atmosphere. Specifically, the collision of cosmic ray particles with atmospheric gases produces neutrons, which then interact with nitrogen-14 to produce C-14. These atoms of C-14 are incorporated into the carbon dioxide in the atmosphere by a series of chemical reactions that occur naturally. Consequently, a small fraction of the carbon in carbon dioxide molecules in the atmosphere is radioactive carbon-14. The amount of C-14 in Earth's atmosphere is close to one atom of C-14 in every trillion atoms of total carbon. (Most of the carbon in Earth's crust and atmosphere is C-12.)

Plants make use of carbon dioxide from the atmosphere in photosynthesis, the process that manufactures the compounds that plants are made of. Therefore, some of the carbon in living plant material is C-14. Many animals eat plants, and therefore some of the carbon in those animals is C-14. Some animals eat other animals, so eventually all animals, too, are made of carbon compounds that contain some C-14. All readers of this book have small amounts of

C-14 in the tissue of our bodies.

The amount of C-14 in the tissues of many living plants and animals has been measured. There is some small variation, but most plants and animals are pretty close to the average. The amount of C-14 in the tissues of living organisms is expressed as the rate of radioactive decay, called the "activity" of the C-14 in the sample, for each gram of total carbon. The C-14 activity is expressed as decays per minute, abbreviated dpm. The currently accepted average for living plants and animals is close to 14 dpm per gram of total carbon.

While a plant is actively growing by photosynthesis, C-14 is being incorporated into the plant tissue. For plants that grow by increments, as many trees do by adding a new growth ring for each growing season, only the growing increment incorporates C-14 into its tissue; increments formed earlier are no longer incorporating C-14. Similarly, as long as an animal is alive and growing or replacing worn out cells with new ones, the animal is incorporating C-14 into its tissue.

When a plant or animal dies, or when an increment in the growth of a tree is no longer actively growing, the amount of C-14 in its tissue decreases over time as the atoms of radioactive C-14 decay into nitrogen-14. The half-life of C-14 is 5730 years. Since the decay rate is proportional to the number of atoms of C-14 in the sample, the activity of the C-14 in the sample also decreases over time. The C-14 activity will decrease to 7 dpm per gram of carbon by the end of one half-life, equal to 5730 years; to 3.5 dpm after two half-lives, equal to 11,460 years; etc.

Most living organisms undergo organic decay after death, and the organism has decayed away completely after a relatively short time. Some materials from living organisms, however, may be preserved under the proper conditions: wood, especially if used in a building, or buried in a deposit that shields the wood from oxygen; bones, especially if entombed in dry rock or in a sarcophagus; arctic animals frozen in permafrost; logs buried in a bog with no oxygen to produce organic decay; and many others.

If material that was once part of a living organism is preserved in such a way that it no longer interacts chemically with the carbon dioxide of the atmosphere, it is possible to determine the age of that material by measuring the amount of C-14 remaining in it. The instruments currently in common use are capable of measuring the activity of C-14 in the sample with good accuracy down to about one one-thousandth of the amount in living organisms. That is the amount of C-14 that would remain after ten half-lives, or 57,300 years. So the method is useful for samples that were once part of a living organism, with ages up to about 50,000 years. A more recent method, using accelerator mass spectrometry instead of decay rate counting, holds out some hope of extending the measureable age range farther back in time, but is still under development.

An hourglass analogy for radiocarbon dating would have a known amount of sand in the top when it began running, analogous to the known amount of C-14 in living organisms today. At any time during the running of the hourglass, we could measure the amount of sand remaining in the top, analogous to the amount of C-14 in a sample at the present time. We measure the rate at which the sand is running through the constriction of the hourglass, analogous to the rate of decay of C-14. Then we could calculate how long the hourglass has been running, analogous to the age of the sample of interest.

Sample selection and sample treatment are critical for the method to yield accurate results. If the amount of C-14 remaining in the sample is very small, even a tiny bit of contamination with modern carbon dioxide (which contains 14 dpm per gram of carbon) will produce a younger age than the correct one. Also, if the sample material has interacted chemically with carbon dioxide from Earth's atmosphere, or with modern carbon dioxide dissolved in water, or some similar interaction during its history, the measurement will produce a younger age than the correct one.

The carbon-14 method of age measurement has proven to be immensely useful, especially in anthropology and archaeology. Measurements on artifacts from Egyptian tombs and monuments provide dates for that ancient history; measurements on the linen in which some of the Dead Sea scrolls were wrapped give us an age for the scrolls; measurements of materials from caves occupied by Stone Age humans provide dates for those periods of history; measurements on tissue from Ice Age creatures found frozen in permafrost in Siberia and Alaska provide dates for the times when those creatures died; etc. Thousands of measurements of C-14 in a variety of samples are carried out every year in laboratories throughout the world.

Validity of the method

How accurate are these age measurements? Are there ways that we can find out?

Procedures for measuring the amount of C-14 in various samples have been refined over the years, and I think we can be confident that the measurements are accurate. The basic method, as described above, is entirely straightforward and reasonable. So the critical question is: How confident can we be that living organisms of the past had the same amount of C-14 in their tissues as living organisms do today? Or, in other words: Is the amount of C-14 being produced in Earth's atmosphere constant, or nearly so, over long periods of time?

One way to check the accuracy of radiocarbon dating for samples from recent millennia is to find samples with a known historical age, and compare that known age with the age as determined by C-14 measurement. That has been done for many samples, ranging in age from relatively recent to artifacts from ancient Egypt that are known to be about 5000 years old. The results show good agreement of C-14 ages with known historical ages.[2]

Another group of samples we might use to test of the accuracy of C-14 ages consists of wood samples from the interior growth rings of living trees, with the age of the sample determined by counting tree rings, and compare that with C-14 ages of wood from those interior growth rings. That has been done. Comparisons were done in the 1960's with samples of various species of trees, including sequoia wood from California, with ages up to 2000 years. This was soon extended farther into the past with wood from the oldest known living trees in North America, the bristlecone pine from Nevada and California, with ages up to 5000 years. Those studies found some short-term fluctuations in the C-14 content of the samples, almost certainly caused by short-term variation in the incidence of cosmic rays onto Earth's atmosphere; the fluctuations are small, amounting to differences in C-14 ages of 50 years or so, some more and some less than the long term average. Trees older than 2500 years were found to have had more C-14 in their living tissue than the current average, increasing for older and older samples to about 10% higher than present values for trees that were alive 5000 years ago.

The technique of tree ring counting has been extended farther into the past than the known ages of any historical artifacts or living trees. The width of each growth ring is affected by climatic conditions during that growth year, and the pattern of variation in the width of growth rings of dead preserved wood samples can be matched with patterns in old living trees. A continuous record can be recovered by "stair-stepping" farther and farther into the past by matching growth ring patterns in many preserved wood samples. By 1970 the comparisons of C-14 ages with wood whose age was known from counting tree rings had been extended from 5000 to 7000 years before the present in bristle cone pine wood, showing short term fluctuations from year to year and from decade to decade, and a continuing level of C-14 in living organisms at about 10% above that in living organisms today.[3,4] The comparison has since been extended to 11,500 years before the present with oak and pine from Germany and Ireland; the differences continue at about 10% for samples that were living from 5000 to 10,000 years ago (counting backwards). From 10,000 years ago to 11,500 years ago we observe a further increase in the C-14 in organisms living at that time to about 12-15% above the levels in living organisms today.[5]

Since we know about those fluctuations and differences between C-14 in modern samples and C-14 in samples up to 11,500 years ago, it is common practice at present to apply a correction to the measured C-14 age of samples within that age range to provide the true age of the sample. The remaining uncertainties in the measurements are not greater than a few percent. That's a pretty strong endorsement of the accuracy of C-14 age measurement.

We have not found any independent method for checking the accuracy of C-14 age measurements by comparison with ages of artifacts or wood older than about 11,500 years. What we have learned about the accuracy of the method for samples younger than that, however, provides confidence that the method gives results that are close to the correct age for samples of any age up to the limits of the measurement at about 50,000 years.

Perspective in publications that promote the view that Earth is a recent creation

Since the C-14 ages of preserved plant and animal remains and many artifacts of human civilization range up to several tens of thousands of years, the reliability of the method has been questioned in publications that promote the view that Earth is a recent creation. We will examine some of the claims of unreliability that have been published.

Radiocarbon dating is discussed in the book *Scientific Creationism*, and the claim is made that the method is unreliable, as follows:

"Despite its high popularity, it involves a number of doubtful assumptions, some of which are sufficiently serious to make its results for all ages exceeding about 2000 to 3000 years, in serious need of revision."[6]

Four such "assumptions" are listed, and we will take a careful look at the fourth one:

"4. <u>The radiocarbon ratio may not have reached a steady state.</u>

"Probably the most important invalid assumption in radiocarbon dating now employed is that the C-14/C-12 ratio is in a steady state with time on a global basis."[7]

What is being referred to in this statement is the production rate of C-14 in Earth's atmosphere, and the decay rate of C-14 in Earth's oceans and living organisms and remains of once-living organisms.

If we were to begin an experiment with a sample that initially contains none of a radioactive isotope, and we turn on a mechanism that produces that radioactive isotope at a constant rate, the number of atoms of the radioactive isotope in the sample will increase as time goes by. As soon as some atoms of the radioactive isotope are produced, however, the process of radioactive decay will also occur, and the number of atoms of the radioactive isotope will be decreased

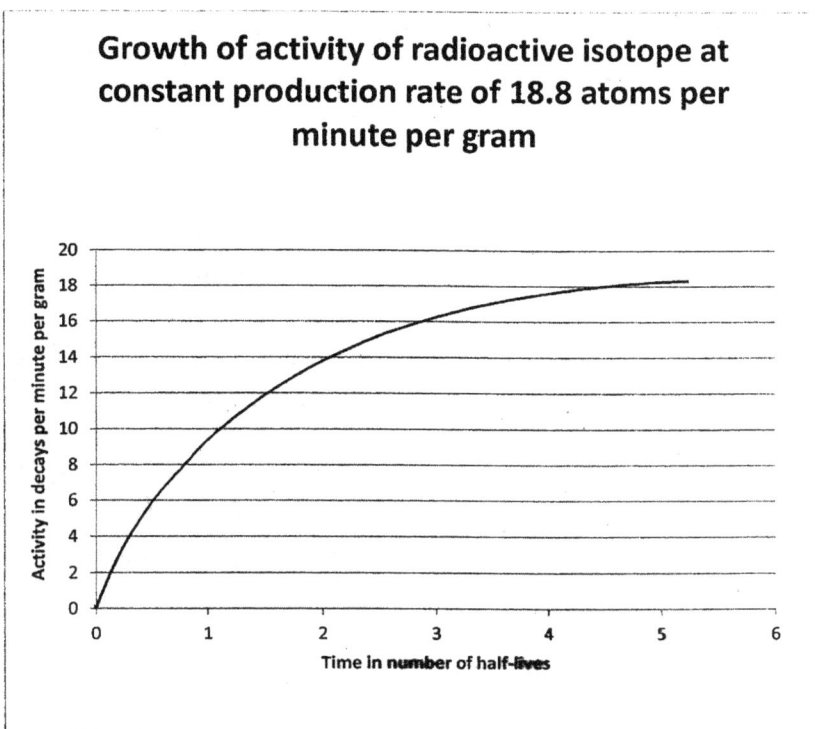

Figure 7. Growth of decay rate at constant production rate.

by that process of radioactive decay. Initially, the rate of decay will be less than the rate of production, but[1] as the number of atoms of radioactive isotope in the sample increases, the rate of decay (for the entire sample) will also increase, until the time that the rate of decay is equal to the rate of production. That condition, when the rate of decay is equal to the rate of production, is called a "steady state," and the decay rate will remain equal to the rate of production thereafter.

A graph showing the growth in the decay rate of the radioactive isotope being produced at a constant rate is shown in Figure 7. The reason for choosing a constant production rate of 18.8 atoms per minute is found in the discussion below. The decay rate of the radioactive isotope in the sample will reach about 99% of its steady state value (the production rate) after 7 half-lives of the isotope, and will be about 99.9% of its steady state value after 10 half-lives of the radioactive isotope.

In describing the radiocarbon dating procedure in his book *Radiocarbon Dating*, Libby published a comparison of his calculated estimates of the production rate of C-14 for the entire Earth with estimates of the decay rate of C-14 for the entire Earth. The values he gives are a production rate of 18.8 atoms of C-14 per gram of total carbon per minute, and a decay rate of 16.1 ± 0.5 decays per gram of total carbon per minute.[1] Because of large uncertainties in the data that he had used for estimating the production rate, Libby concluded that the production rate and decay rate of C-14 could be considered to be nearly equal, that is, the C-14 system is in a "steady state."

Libby did not include an estimate of the amount of uncertainty of his production rate in his initial publication of these data, but he did list the sources of the data he used for his calculation, and those data have large uncertainties. A more technical and detailed presentation of those uncertainties is found in Appendix E.

Evaluation of the recent Earth claim

The discussion of C-14 production and decay rates in *Scientific*

Creationism calls attention to the difference between the estimated production rate and the decay rate of C-14, as published by Libby, and suggests that those data support the conclusion that the C-14 system has not yet reached a steady state condition, and that those data support the belief that Earth is a recent creation.

An article that had been published in the *Creation Research Society Quarterly* is referred to in a footnote in *Scientific Creationism*; that article presents the same argument in more detail.[8] The article in *CRS Quarterly* included a graph depicting the growth of the decay rate of a radioactive isotope that is being produced at a constant rate; like the graph above, it used the rate of 18.8 atoms per minute, the rate estimated by Libby for production of C-14. The argument in *CRS Quarterly* is based on the following proposed conditions:

1. There was no C-14 in the tissues of living organisms when the production of C-14 began in Earth's atmosphere.

2. The rate of production of C-14 in the atmosphere has been constant since that process began.

3. The values published by Libby, namely, a present decay rate of 16.1 decays per gram of total carbon per minute and a constant production rate of 18.8 atoms of C-14 per gram of total carbon per minute for the entire Earth, are the correct values.

4. The uncertainties in Libby's estimates can be ignored, and the values listed in point 3 above are the precisely correct values.

5. The reason the estimated worldwide decay rate is less than the estimated production rate is that the decay rate has been increasing toward equilibrium with the production rate, but has not yet reached that equilibrium value.

As you can see from the graph in Figure 7, the measured decay rate of 16.1 atoms of C-14 per minute per gram of total carbon falls on that graph at a bit less than 3 half-lives after production began; the half-life of C-14 is 5730 years, so the suggestion in the *CRS Quarterly* is that production of C-14 on Earth began 15,000 years ago. If the suggestion in the *CRS Quarterly* is correct, the rate of decay of C-14 in living organisms on Earth would have been less in the past than it is today, as indicated in the graph in Figure 7.

Is there some way that we might check whether that suggestion has any merit, other than comparing estimated production rate and decay rate? Yes, there is.

Based on the ages of artifacts with known historical age and wood whose age has been determined by tree ring dating, we can determine how much C-14 was present in the sources of those samples when they were alive. Using the graph shown above, we can compare the suggestion from the article in the *CRS Quarterly* with the amount of C-14 that actually was present in those organisms when they were alive.

The line in the graph that shows the changing decay rate with time suggests that organisms living on Earth 5000 years ago would have had about 10% <u>less</u> C-14 in their tissues than we find in living organisms today if the conjecture in the article in the *CRS Quarterly* is correct; in actual fact, the measurements on samples of that age show that they had about 10% <u>more</u> C-14 than today. The graph suggests that organisms living on Earth 8000 years ago would have had about 20% <u>less</u> C-14 than living organisms today; in actual fact, tree ring samples of that age had about 10% <u>more</u> than today. If we go back to the oldest tree ring data available, 11,500 years ago, the graph suggests that the amount of C-14 in living organisms would have been about 50% <u>less</u> than today; in actual fact, tree ring samples of that time actually had 10-15% <u>more</u> than today. The difference between the suggestion in the *CRS Quarterly* article and the actual observations of samples of God's world is large. It is obvious that the suggestion in the *CRS Quarterly* and in *Scientific Creationism*, namely, that the radiocarbon ratio may not have reached a steady state, is contradicted by the actual data.

If truth be known, the suggestion that the worldwide decay rate of C-14 has been increasing and has not yet reached a steady state with C-14 production had already been discredited by the data published in Damon (ref [3] above) and by Seuss (ref [4] above), before *Scientific Creationism* was published (ref [6] above).

Our understanding today

Libby's estimates of the worldwide production rate and decay rate of C-14 were published in the early 1950's. What have we learned since then?

A great deal of data has been collected since 1950. We know that the rate of cosmic ray bombardment of the Earth's atmosphere, and therefore the rate of C-14 production, are affected by variations in the strength of Earth's magnetic field, and by variations in solar activity, that is, sunspot and solar flare cycles. Variations in solar activity are short term cycles, and these had been detected and mapped for the past 7000 years by 1970.[3,4] The data on samples whose age is determined by tree ring counting have also shown that the C-14 content of living organisms was higher than it is today during several millenia prior to 500 B.C.; this undoubtedly reflects an earlier time when the production rate of C-14 was higher than it is today.

Additional data have also been collected regarding the neutron flux (number of neutrons in a given volume of space at any moment) in the atmosphere, leading to better estimates of C-14 production rate than was available to Libby in 1950. The neutron flux varies with altitude in the atmosphere, and also with latitude relative to the magnetic field surrounding Earth. A revised rate, based on data over three solar cycles, was published in 1970; 2.2 ± 0.4 atoms of C-14 per square centimeter of Earth's surface per second.[9]

Additional data have also been collected regarding the worldwide activity (decay rate) of C-14 in the carbon-containing materials participating in the "carbon cycle," which includes the atmosphere, biological organisms on Earth, ocean water, and recently deposited sediments. In particular, recent studies have found that the amount of C-14 found in sediments is higher than earlier estimates, leading to higher estimates of worldwide C-14 activity than earlier estimates. A couple of estimates published in 1990 give values of 2.1 and 2.38 ±

0.13 decays per cm^2 of Earth's surface per second.[10] These values are well within the uncertainty limits of the estimated production rate of 2.2 ± 0.4 atoms of C-14 per cm^2 per second, amply justifying the conclusion that the decay rate of C-14 on Earth has reached a steady state, equal to the production rate.

There is another valuable bit of information about the long term rate of cosmic ray bombardment of Earth's atmosphere that has been derived from samples of the moon returned by the Apollo space flights. Since the moon has no atmosphere, the cosmic ray particles bombard the moon's surface directly, and produce various nuclear reactions in the materials at and near the moon's surface. Some of the products of these bombardments are isotopes of radioactive materials. The concentrations of several different isotopes, covering a range of half-lives, have been measured in lunar samples returned by the Apollo flights. The probabilities (known technically as "cross-sections") for the production of various isotopes by cosmic ray particles have been measured with the use of particle accelerators (cyclotrons, etc.) in Earth's laboratories. The ratios of radioactive isotopes from the lunar samples serve as a test for long-term variations in the rate of cosmic ray bombardment on the moon, and the results indicate that the rate of cosmic ray incidence on the moon has been very nearly constant at or near its present rate for at least the past one million years.[11] Since moon and Earth are near neighbors in space, the same conclusion holds for Earth. Similar results had been reported earlier, based on studies of isotopes produced in meteorites by bombardment with cosmic rays.[12]

If truth be known, it seems safe to conclude that short-term fluctuations in the rate of cosmic ray bombardment of Earth's atmosphere may result in small short-term variations in C-14 production in Earth's atmosphere, but the long term rate of C-14 production has been nearly constant for a period of time longer than many half-lives of C-14. Consequently, the production and decay of C-14 on Earth has reached a steady state long ago, and the radiocarbon dating method gives results that are close to correct.

References

[1] Willard F. Libby, *Radiocarbon Dating.* (Chicago, University of Chicago Press, 1955), 7.

[2] Libby, *Radiocarbon*, 10.

[3] Paul E. Damon, A. Long and D.C. Grey, "Arizona radiocarbon dates for dendrochronologically dated samples" in *Radiocarbon Variations and Absolute Chronology, Nobel Symposium 12*, Ingrid U. Olsson, Ed., (New York: John Wiley & Sons, 1970), 615-8.

[4] H.A. Suess, "Bristlecone-pine calibration of the radiocarbon time-scale 5200 B.C. to the present" in *Radiocarbon Variations*, 303-11.

[5] Minze Stuiver, Paula J. Reimer and Thomas F. Braziunas, "High-Precision Radiocarbon Age Calibration for Terrestrial and Marine Samples," *Radiocarbon* 40 (1998), 1127-51.

[6] Henry M. Morris, Ed., *Scientific Creationism.* (San Diego: Creation-Life Publishers, 1974), 162.

[7] Morris, *Creationism*, 164.

[8] Robert L. Whitelaw, "Radiocarbon Confirms Biblical Creation," *Creation Research Society Quarterly* 5 (1968), 78.

[9] R.E. Lingenfelter and R. Ramaty, "Astrophysical and geophysical variations in C-14 production," in *Radiocarbon Variations*, 513-35.

[10] Robert S. Sternberg, "Radiocarbon fluctuations and the geomagnetic field," in *Radiocarbon After Four Decades*, R.E. Taylor, A. Long, and R.S. Kra, Eds., (New York: Springer Verlag, 1992), 102.

[11] L.A. Rancitelli, R.W. Perkins, W.D. Felix and N.A. Wogman, "Lunar surface processes and cosmic ray characterization from Apollo 12-15 lunar sample analyses" in *Proceedings of the Third Lunar Science Conference.* (Cambridge, MA: MIT Press, 1972), 1681.

[12] J.R. Arnold, M. Honda and D. Lal, "Record of Cosmic-Ray Intensity in the Meteorites," *Journal of Geophysical Research* 66 (1961), 3519.

22 More on Carbon-14

In the preceding chapter we examined the claim that "the radiocarbon ratio may not have reached a steady state," and we concluded that the suggestion has no merit. We should consider two additional claims in *Scientific Creationism* related to what are called "doubtful assumptions" for radiocarbon dating.

Total carbon in the carbon reservoir

We next consider the suggestion in *Scientific Creationism* that:

"The amount of natural carbon may have varied in the past,"[1]

As we observed in the previous chapter, the radiocarbon dates are in good agreement with the known ages of historical samples and the tree ring ages of ancient wood for the past several millennia. We used that information to support the conclusion that the C-14 system has reached a steady state, with the production rate equal to the decay rate for the Earth as a whole. However, if the amount of C-14 in Earth's atmosphere has been increasing from zero at some time in the recent past, and if the amount of total carbon that is interacting with the carbon-14 in the atmosphere has been increasing at approximately the same rate, it would appear to us at present that the ratio of C-14 to total carbon in living organisms has been constant, or nearly constant, throughout that history. In other words, the radiocarbon dates would agree with the historical dates, even though the amount of C-14 in Earth's crust and atmosphere has been increasing, because the amount of total carbon has been increasing at the same rate. This possibility was developed more fully in the book *Adam When?*[2]

Estimates have been made of the total amount of carbon in the

Earth's crust and atmosphere that is interacting chemically with the C-14 in Earth's atmosphere in the "carbon cycle." This would include all living organisms, various carbonates in ocean water, decayed and decaying animal and vegetable matter in the top several feet of soil on Earth's surface, and some of the carbonates carried in stream water and near-surface groundwater.

The suggestion in *Adam When?* leads the reader through a series of computations, starting with the following assumed (not necessarily justified) conditions:

1. Carbon-14 began to be produced in Earth's atmosphere 13,000 years ago,

2. The rate of production of C-14 has been constant at 2.5 atoms of C-14 per square centimeter of Earth's surface per second since that time,

3. The amount of total carbon on Earth at the present time is 8.79 grams of total carbon per square centimeter of Earth's surface,

4. The rate of decay of C-14 for the entire Earth is now at about 80% of the production rate.

[Note: the production rate of C-14 in #2 above is an estimate from Lingenfelter.[4]]

The computations then make use of the C-14 content of samples of living organisms as reported in the current scientific literature[5] to determine what the rate of decay of C-14 would have been 7000 years ago, and how much total carbon would have been present in the carbon cycle 7000 years ago, if the conditions stated above were correct. The computation yields a decay rate of 0.85 decays of C-14 per second per cm² of Earth's surface, and total carbon of 3.73 grams of carbon per cm² of Earth's surface 7000 years ago. (The beginning time of 7000 years ago was presumably chosen because the oldest tree ring samples available at the time the book was written were 7000 years old.)

The increase in total carbon from 3.73 grams per cm² to 8.79 grams per cm² for the entire Earth's surface since 7000 years ago would require the addition of 28,000 billion tons of carbon to the Earth. That would be an average of 4 billion tons per year, equivalent to adding 14.8 billion tons of carbon dioxide per year. That's a lot of carbon! Is that a reasonable suggestion? Where would that much carbon come from over the past 7000 years?

The discussion in *Adam When?* considers potential sources of additional carbon,[6] and makes reference to an article on the carbon in the carbon cycle that was published in the monthly magazine *Scientific American*. That article reports an estimate that the amount of carbon dioxide in Earth's atmosphere has been increasing at an average rate of 7.9 billion tons per year during the century from 1860 to 1959.[7] Actually, about 2 billion tons of that total came from the disturbed soil of recently cleared agricultural land, which was already part of the carbon reservoir. So the amount of new carbon dioxide added to the atmosphere during that time was an average of 6 billion tons per year, all of it from the burning of fossil fuels - coal, natural gas, and crude oil.

The discussion in *Adam When?* notes that the annual increase of 7.9 (actually, 6) billion tons of carbon dioxide reported in *Scientific American* is somewhat less than the annual 14.8 billion tons required to support the suggestion that the total carbon in the carbon cycle has increased at the same rate as the proposed increase in C-14, but suggests that it is "of the same order." Then the claim is made that an annual increase of 7.9 billion tons over the century 1860-1959 justifies the proposal that an annual increase of 14.8 billion tons over the past 7000 years is reasonable:

"It is possible from the secular evidence to show that an increase from an average of 3.7 grams of per cm^2 of carbon in the carbon cycle in 4990 B.C. to 8.8 grams per cm^2 at the present time is quite realistic."[8]

Is that a valid claim? Is that a reasonable conclusion based on the calculations presented?

The only suggested source for adding those billions and billions of tons of new carbon dioxide to the carbon reservoir is the burning of fossil fuels. Humans have been burning a significant amount of fossil fuels only since the beginning of the Industrial Revolution. The amount burned before the century from 1860 to 1959 (the century examined by Plass in *Scientific American*) is very small indeed. So we are left with less than half of the new carbon required by the suggestion in *Adam When?* for the century ending in 1959, and no potential source whatever for the immense amounts of new carbon needed during the 6,900 years before that.

If truth be known, there is no scientific basis for thinking that the total amount of carbon in the carbon reservoir has undergone any significant change for the past very long time, except for the recent burning of fossil fuels. Therefore, the claim in *Adam When?* and the suggestion in *Scientific Creationism* that there has been a significant increase in the total amount of carbon interacting with the C-14 in Earth's atmosphere over the past several thousand years is refuted by the scientific information available to us.

Living organisms with reduced C-14 levels

The discussion in *Scientific Creationism* also suggests that C-14 ages might be in error because

"Many living systems are not in equilibrium for C-14 exchange.
"The C-14 method assumes the standard C-14/C-12 ratio applies to all living organisms at the time of death. That this is not correct has been shown in many instances. For example, it has been found that the shells of living mollusks may show radiocarbon ages of up to 2300 years."[9]

The reference given to support this claim is an article entitled "Radiocarbon Dating: Fictitious Results with Mollusk Shells," published in *Science*.[10] So, what is going on here?

The mollusks in question are fresh-water snails, taken from stream water. The bedrock of the drainage basin of the stream is limestone, which consists mostly of calcium carbonate. Since the limestone was deposited long ago, and the half-life of C-14 is only 5730 years, there is no C-14 in the limestone at the present time. The ratio of C-14 to total carbon in the carbon compounds that are dissolved in the stream water is well below the ratio in the atmosphere and in most living plants and animals because some of the dissolved carbon compounds come from the weathering of the limestone. The snails get the material for building their shells from the water they live in, so the ratio of C-14 to total carbon in their shells is the same as that in the carbon compounds dissolved in the stream water, which is from mixed sources. The ratio of C-14 to total carbon in their shells gives them an apparent age of 2300 years, but the decreased level of C-14 in these shells is due to the C-14 deficiency in the carbon compounds dissolved in the stream water, and not to the passage of time.

So the reason for the C-14 deficiency in the shells of these living mollusks is well understood. The ratio of C-14 to total carbon in their shells is, in fact, in equilibrium with the ratio of C-14 to total carbon in the environment from which they derive material to build their shells.

It is also well known that recent growth in trees that are near and downwind from industrial plants that burn fossil fuels (coal, natural gas, etc.) is depleted in C-14, thus giving a fictitious C-14 "age." The carbon dioxide that those trees use in photosynthesis includes some carbon dioxide produced by burning those fossil fuels, which contain virtually no C-14.

The researchers who perform radiocarbon age measurements read the scientific journals, and often write the articles in those journals. They are aware of the kinds of pitfalls that await those who are less than well informed, and they understand how to choose samples with care. If truth be known, there is no scientific foundation for the claim in *Scientific Creationism* that fictitious C-14 ages in the freshwater mollusks, and any similar cases, would make all radiocarbon age measurements suspect.

References

[1] Henry M. Morris, Ed., *Scientific Creationism*, (San Diego: Creation-Life Publishers, 1974), 162.

[2] Harold Camping, *Adam When?* (Alameda, CA: Frontiers for Christ, 1974).

[3] Camping, *Adam When?*, 194.

[4] R.E. Lingenfelter, "Production of Carbon 14 by Cosmic-Ray Neutrons," *Reviews of Geophysics* 1 (1963), 35.

[5] H.A. Suess, "Bristlecone-pine calibration of the radiocarbon time-scale 5200 B.C. to the present" in *Radiocarbon Variations and Absolute Chronology, Nobel Symposium 12*, Ingrid U. Olsson, Editor. (New York: John Wiley & Sons, 1970).

[6] Camping, *Adam When?*, 227.

[7] Gilbert N. Plass, "Carbon Dioxide and Climate," *Scientific American* 201, No. 1 (July 1959), 41-7.

[8] Camping, *Adam When?*, 227.

[9] Morris, *Creationism*, 162.

[10] M.L. Keith and G.M. Anderson, "Radiocarbon Dating: Fictitious Results with Mollusk Shells," *Science* 141 (16 August 1963), 634.

23 Neutrons, Uranium Ores, and Lead Isotopes

In the discussion of the uranium-lead methods for measuring the ages of rocks, the book *Scientific Creationism* claims that the method could be rendered unreliable if the rock deposit has been exposed to a large number of neutrons. The details of that suggestion are expressed as follows:

"An even more important phenomenon by which these balances [of components of the uranium-lead system] can be upset is that of 'free neutron capture,' by which free neutrons in the mineral's environment may be captured by the lead in the system to change the isotopic value of the lead. That is, Lead 206 may be converted into Lead 207, and Lead 207 into Lead 208 by this process. It is perhaps significant that Lead 208 usually constitutes over half of the lead present in any given lead deposit. Thus, the relative amounts of these 'radiogenic' isotopes of lead in the system may not be a function of their decay from thorium and uranium at all, but rather a function of the amount of free neutrons in the environment.

"That this is a serious problem has been shown by Melvin A. Cook, who has analyzed two of the world's most important uranium bearing ores (e.g., in Katanga and Canada) with this in view. These ores contain no Lead 204, so presumably no common lead. They also contain little or no Thorium 232, but do contain significant amounts of Lead 208. The latter could therefore have come neither from common lead contamination, nor from thorium decay, and so must have been derived from Lead 207 by neutron capture."[1]

The reference given for the claims and calculations of Melvin A. Cook is to the book *Prehistory and Earth Models*.

Remember that the procedure for measuring the age of a rock using uranium-238/lead-206 was presented in some detail in Chapter 17 and Appendix C, so there is no need to repeat that here. We should remind ourselves, however, that the amount of Pb-204 in the sample is important to the procedure, because Pb-204 is not the product of radioactive decay of any isotope presently found on Earth, and therefore the amount of Pb-204 in the sample remains constant throughout the history of the rock. The amount of lead-204 in the sample provides the basis for determining how much of the Pb-206 or Pb-207 in a uranium-bearing sample was present when the rock was formed, and how much was produced by decay of radioactive uranium isotopes.

The suggestion is made in *Prehistory and Earth Models* that the Earth, or some regions of the Earth, may have experienced a bombardment by neutrons at some undetermined time in Earth's history. This bombardment may have taken place, according to the suggested scenario, by neutrons with the same range of energies as are observed from the fission of uranium isotopes. The further claim is then made that:

"there are several well documented examples that seem to demonstrate the reality of this scheme [that is, that the ratios of lead isotopes has been altered by bombardment with neutrons]. Consider, for example, the uranium ore body of Shinkolobwe, Katanga (see Faul 1954). In this ore quite generally lead-204 is zero as also is thorium. Yet there is present in this ore some lead-208! Where did it come from? The absence of lead-204 implies that there is no original lead [usually referred to as "common lead"] in this ore; apparently all of the lead is radiogenic. On the other hand, since there is no thorium either, the lead-208 could not have come from thorium decay. It evidently, therefore, had to come from the (neutron in, gamma ray out) reaction 'lead-207 + neutron yields lead-208 + gamma ray.'"[2]

Later:

"As another example, consider the Martin Lake ore of the Canadian Shield. ... There was found an average of 0.53% lead-208 in this ore but enough thorium to account for less than 1% of this lead-208. Apparently, practically all of it, therefore, came from the (neutron in, gamma ray out) transmutation of lead-207."[3]

If truth be known, however, the assertion that these ores contain "zero" lead-204 is incorrect!

What was the source of the data on the lead isotopes in those uranium ores from Shinkolobwe and Martin Lake that was used in *Prehistory and Earth Models*? How could incorrect data have been introduced into the argument presented in that publication?

The source for the isotope data in *Prehistory and Earth Models* is identified as the book *Nuclear Geology*.[4] The isotope data presented in *Nuclear Geology* were copied from earlier professional papers: the data for Shinkolobwe from Alfred O. Nier;[5] Collins, Farquhar, and Russell;[6] and Ehrenberg,[7] and the data for Martin Lake from Collins et al.;[6] and Kerr and Kulp.[8] When we consult the original papers we find that the amount of lead-204 was, in fact, measured in several of the samples. Results of the measurement of lead-204 were reported for four of the five samples of Shinkolobwe ore analyzed by Collins et al., and for one of the eight samples of Shinkolobwe ore reported by Nier, and "less than" values were reported for three of those eight samples. Results of lead-204 measurements were reported for two of the twelve samples of Martin Lake ore analyzed by Collins et al., with an additional sample reported as "trace." The concentration of lead-204 is certainly not "zero" in those samples.

For the remainder of the samples, the amount of lead-204 was reported in the original papers and in *Nuclear Geology* as (----). That "dash," however, does not mean "zero." Nier states that:

" …. In a number of samples where the abundance of lead-204 was very low no attempt was made to measure the amount."[9]

The discussion of uranium/lead methods in *Nuclear Geology* also includes the statement

"Common lead in a uranium mineral is indicated by the presence of lead-204 and, in thorium-free districts, by the presence of lead-208."[10]

Since thorium-232 is the parent of lead-208, the amount of lead-208 in the sample does not change with the passage of time in samples containing no thorium. Therefore, the measurement of the amount of lead-208 in those samples can serve as the basis for determining the initial amount of lead isotopes in the sample when the rock was formed. It wouldn't be necessary to measure the amount of lead-204 in those samples. Since the abundance of lead-208 is generally about 37 times the abundance of lead-204 in rocks in the Earth's crust, it is much easier and more accurate to measure the amount of lead-208 in samples containing very little original lead than to attempt to measure the much smaller quantities of lead-204.

The claim in *Prehistory and Earth Models*, namely, that the uranium ore at Shinkolobwe and at Martin Lake must have been bombarded by neutrons to produce the lead-208 found in them, depends on the assertion that "In this ore quite generally lead-204 is zero ... "[11] Since that assertion is incorrect, the claim of bombardment by neutrons is incorrect as well.

One has to wonder how the incorrect claim that "these ores contain no lead-204" came to be published in *Prehistory and Earth Models* and in *Scientific Creationism*. I can only guess, but I offer the following: The passage in *Prehistory and Earth Models* refers to a secondary publication (Faul), in which the data were copied from the earlier papers with only (----) values for lead-204 for many of the samples. The bibliography found in *Prehistory and Earth Models* and the bibliography and footnotes in *Scientific Creationism* make no mention of the original papers by Nier, by Collins et al., by Ehrenberg, or by Kerr and Kulp. My guess is that the original papers were not consulted, that the comment in *Nuclear Geology* to the effect that lead-208 can be used as an indicator of common lead in thorium-free deposits was overlooked, and that it was mistakenly assumed that the (----) values in *Nuclear Geology* meant "zero."

The scenario in *Prehistory and Earth Models* is fatally flawed in another way, as well. If there was no common lead in the ore when it was deposited, then there was no lead-206 that could be converted to lead-207, nor any lead-207 to be converted to lead-208. Only as lead-206 was formed by radioactive decay of U-238, and lead-207 by radioactive decay of U-235, could any lead-208 be formed at all, and thus the ore would already be very old before even a very small amount of lead-208 could be formed by neutron bombardment.

Another consideration

Let's suppose for a moment that there might have been a massive neutron bombardment of the Earth, including its uranium ore deposits, at some time in the past. Might there be some way of finding out about that? Might we be able to find evidence, other than hypothetical changes in lead isotope ratios, that such an event actually occurred, or that it did not?

Yes, there are ways to find out.

Any neutron bombardment would interact with whatever elements were present where such a bombardment occurred. For our evaluation of this case, let us consider reactions of fast, energetic neutrons, such as those suggested in *Prehistory and Earth Models*, on the uranium in uranium-rich rocks. Such a bombardment would induce an appreciable amount of fission of both uranium-238 and uranium-235 atoms. The fission products, as we well know from our experience with modern nuclear reactors, are radioactive. Some of them have long half-lives; technetium-99, an isotope produced in 6% of such fissions, has a half-life of 213,000 years. If such a bombardment had taken place within the past million or so years (about 5 times the half-life of technetium) we should find an appreciable amount of technetium-99 in those uranium deposits at the present time. But it isn't there, so we can confidently conclude that there has not been any massive fast neutron bombardment of those uranium deposits, at least not within the past million years.

That's just one example. We could readily add more, and we could also consider reactions of neutrons with other elements in the deposit to see what the products would be, and then find out whether those products are found there or not. But this one example should suffice to convince us that the proposed neutron bombardment has not occurred.

Results of a neutron bombardment on elements other than lead were not considered in *Prehistory and Earth Models*, nor in *Scientific Creationism*. If they had been, as would have been proper and expected for any new and unusual proposal in science, it should have become apparent that the suggested scenario of alteration of lead isotope ratios by neutron bombardment was not a valid proposal.

And so, if truth be known, the entire suggested scenario in *Prehistory and Earth Models*, claiming evidence that a neutron bombardment has altered lead isotope ratios in some uranium ore deposits, amounts to exactly nothing, because it is based on the incorrect assertion that these uranium ores contain no lead-204, and because it is inconsistent with other evidence. The claim in *Scientific Creationism* that it has been shown in *Prehistory and Earth Models* that such a neutron bombardment presents a serious problem with uranium-lead isotope age measurements, therefore, is also without merit.

You may hold to the view that Earth is a recent creation for other reasons, if you wish, but the possibility of neutron bombardment of uranium ores has been examined and shown to be mistaken, so that suggestion does not provide support for the view that Earth is a recent creation.

References

[1] Henry A. Morris, Ed., *Scientific Creationism*, (San Diego: Creation-Life Publishers, 1974), 141.

[2] Melvin A. Cook, *Prehistory and Earth Models*, (London: Max Parrish, 1966), 54.

[3] Cook, *Prehistory*, 55.

[4] Henry Faul, Ed., *Nuclear Geology, A Symposium*, (New York: John

Wiley & Sons, and London: Chapman & Hall, Ltd, 1954).

[5]Alfred O. Nier, "The Isotopic Constitution of Radiogenic Leads and the Measurement of Geological Time, II," *Physical Review* 55 (1939), 153-63.

[6]C.B. Collins, R.M. Farquhar, and R.D. Russell, "Isotopic constitution of radiogenic leads and the measurement of geological time." *Bulletin of the Geological Society of America*, 65 (January 1954), 1-22.

[7]H. Fr. Ehrenberg, "Isotopenanalysen an Blei aus Mineralen" (Isotope analyses of lead from minerals). *Zeitschrift fur Physik* 134 (1953), 317-33.

[8]P.F. Kerr and J.L. Kulp, "Pre-Cambrian uraninite, Sunshine Mine, Idaho," *Science* 115 (1952), 86-8.

[9]Nier, "Isotopic Constitution," 156.

[10]Faul, *Nuclear Geology*, 283, 293.

[11]Cook, *Prehistory*, 54.

24 Excess Argon-40

Many publications promoting the view that Earth is a recent creation have included claims that the results of scientific study actually support that view, rather than the old-Earth conclusions that are accepted by nearly all professional scientists. That is the stance, for example, of the movement that calls itself "scientific creationism." An internet web site, www.answersingenesis.org, also publishes such claims.

The flip side of "scientific creationism" is characterized by efforts to discredit the reliability of the scientific methods of age measurement that support the "old-Earth" view. In this chapter we will consider a claim that the K-40/Ar-40 method of measuring ages of rocks is unreliable because some "excess" Ar-40 is found in very young deposits of volcanic rocks. This claim is published, for example, in the book *Scientific Creationism*. Following quotations from the scientific literature reporting "excess" Ar-40 found in recent lava samples from Hawaii, we read:

> "The creationist does not question the fact that the anomalously high ages of the lava rocks noted above may well be due to incorporation of excess argon at the time of formation. Again, however, he points out that if this is known to have happened so frequently in rocks of known age, it probably also happened frequently in rocks of unknown age. Since there is no way at all to distinguish Argon 40 as formed by unknown processes in primeval times and now dispersed around the world, from radiogenic Argon 40, it seems clear that potassium-argon ages are meaningless insofar as *true* ages are concerned."[1]

So let us take a more careful look at the articles on the subject that have been published in the scientific literature to help us understand what is going on.

One such study reports measurements of Ar-40 in lava that was extruded below sea level on the flanks of the Kilauea volcano on the island of Hawaii.[2] The samples were dredged from several different depths, ranging from 1400 to 4680 meters (4592 to 15,350 feet) below sea level. Lava that is extruded under water forms a structure known as "pillow lava" because the lava is cooled very quickly in the water and forms pillow-shaped lobes. When lava is cooled quickly it forms a solid known as a "glass," in which atoms of the elements that were in the molten magma are in random arrangement, and not in the regular geometric pattern that we find in mineral crystals.

Measurements done on those pillow lava samples found an appreciable amount of argon-40 present in some (but not all) of the samples, although the lava was recent, not more than a couple of hundred years old. If one assumed that there was no Ar-40 in the lava when it was extruded from the volcano, measurement of the age of the lava by the K-40/Ar-40 method would give results ranging up to 22 million years, which is obviously incorrect.

So there are some questions that we should seek answers to:

1. Where might the Ar-40 have come from, if not from radioactive decay of the K-40 in the rock sample?

Answer:

a. Almost certainly not from the air, although about 1% of the volume of Earth's atmosphere is argon. The lava never contacted the atmosphere in being extruded and cooled, and only the outermost surface of the pillow contacted sea water.

b. Argon is a gas at the temperatures involved. Like other gases, it can be dissolved in liquids if it is kept under pressure. Like the carbon dioxide in carbonated soft drinks, argon would form bubbles and escape from the liquid if the pressure were lowered to atmospheric pressure at sea level. These lava samples, however, were extruded below sea level, and therefore cooled and solidified under pressure, where the argon is likely to remain dissolved in the molten magma.

c. Very likely, the Ar-40 in those samples had been dissolved in the molten magma below ground, was carried along when the magma was extruded into the surrounding sea water below sea level, and was trapped in the cooled solid.

2. Why was the argon trapped in the pillow lava?

Answer:

a. The rapid cooling of the lava upon coming into contact with sea water, coupled with the high pressure at depth, did not allow the argon to escape as bubbles.

b. The glassy nature of the lava, with atoms in random order, permitted the argon atoms to be trapped in the glass. In a crystalline igneous rock, with atoms arranged in the three-dimensional patterns of mineral crystals, there would be no room within the mineral crystal for argon atoms.

3. Is there some additional reason to believe that the argon did not come from the atmosphere? After all, there was certainly some air from Earth's atmosphere dissolved in the sea water into which the lava was extruded, and perhaps even in the molten magma far below Earth's surface.

Answer: Yes, there is evidence that the excess argon did not come from the atmosphere. The atmosphere is approximately 1% argon. At the present time, 0.337 % (less than 1%) of the argon atoms in the atmosphere are Ar-36; even fewer, 0.063 % are Ar-38; and the remaining more than 99% are Ar-40. It is standard procedure in potassium-argon dating of rocks to make a correction for air that might have leaked into the evacuated mass spectrometer during the measurement, based on the amount of Ar-36 found in the sample, if any. That correction would also have removed from consideration any Ar-40 that might have entered the pillow lava from atmospheric argon.

Another study, undertaken at about the same time, was done on samples from the flanks of Kilauea ranging from 500 to 5000 meters (1640 to 16,400 feet) below sea level; Ar-40 contents of those samples would normally be associated with ages ranging up to 43 million years.[3] In that study, analyses were also performed on samples that were taken from different locations in a single "pillow," ranging from near the rim to 11 centimeters into the interior of the pillow. Those nearest the rim contained higher amounts of argon-40, indicating an "age" of 43 million years, while those at 8 and 11 centimeters from the rim contained considerably less argon-40, indicating "ages" of 1.0 and 1.5 million years. These observations indicate that the lava near the rim cooled most rapidly, forming glassy lava that trapped more argon gas, and the interior of the pillow cooled a little bit more slowly, producing less glassy rock and trapping less argon.

A third publication, also at about the same time, reported excess Ar-40 in a lava flow from an eruption of the Hualalai volcano that had occurred in 1801-1802.[4] The samples for the study were "xenoliths," consisting of minerals that solidify from molten magma at high temperatures. Xenoliths are formed when high-temperature minerals crystallize deep underground as the body of magma cools to below their crystallization temperature, and then the solid xenoliths are carried to the surface by the mostly liquid magma in a volcanic eruption. After the remaining magma cools and solidifies, the xenoliths are easily recognized as rounded, crystalline inclusions, differing from the surrounding fine-grained lava.

These xenoliths were found to contain excess Ar-40, that is, argon that was present when the lava was extruded. How did that argon become trapped in those samples?

In the usual technique of K-40/Ar-40 age measurements, the sample is heated to melting, thus releasing the argon gas. In the case of these xenoliths, nearly all the Ar-40 was also released by crushing the sample and grinding it very fine, without heating it to melting. Microscopic examination of the crushed fragments indicated that the trapped Ar-40 in these xenoliths was located primarily in small vesicles (called "bubblets" by the authors) which contained some fluid and gas, while there was little or no Ar-40 in the solid portion of the crystals. The conclusion is that the Ar-40 that had been dissolved in the magma deep underground was incorporated into the xenoliths as they solidified from the melt, but in fluid and gaseous pockets, not within the mineral crystal structure.

An entry on the website www.answersingenesis.org reports that some samples of recent lava from Mt. Ngauruhoe in New Zealand also contained excess Ar-40.[5] The amounts of Ar-40 in various samples ranged from too little to measure, giving an apparent age of zero, to an amount that would give an apparent age of 3.5 million years.

This report from New Zealand volcanoes is not bringing us new information; thirty years earlier a paper was published in the scientific literature on excess Ar-40 found in lava from the Auckland volcanic field.[6] Such results would be consistent with more rapid cooling of some samples than others, resulting in varying amounts of glassy material in the samples. As discussed above, it is not surprising to find excess Ar-40 in glassy material in recent lava flows.

The discussion of excess argon found in recent Hawaiian lava flows in *Scientific Creationism* states that:

"… it seems clear that potassium-argon ages are meaningless insofar as *true* ages are concerned."[7]

The article in www.answersingenesis.org on samples from lava flows on Mt. Ngauruhoe in New Zealand concludes:

"We know the true ages of the rocks because they were observed to form less than 50 years ago. Yet they yield 'ages' up to 3.5 million years which are thus false. How can we trust the use of this same 'dating' method on rocks whose ages we don't know?"[8]

Do the results reported in the papers described above mean that all measurements of ages by K-40/Ar-40 are unreliable? Certainly not. It's a matter of sample selection. The mechanisms for trapping Ar-40 from the molten magma in glassy lava flows and in xenolithic nodules in lava flows are well understood. The fact that the scientific community has been alerted to this potential problem means that care must be exercised in selecting samples for age determination by K-40/Ar-40 methods, but measurements of ages of crystalline igneous rocks such as granite, gabbro, etc. would certainly not be expected to suffer from error because of any trapped excess Ar-40.

You may choose to reject the results of age measurements of rocks using the K-40/Ar-40 method for other reasons, if you choose to do so. But, if truth be known, the observation of trapped excess Ar-40 in recent lava flows is a warning about sample selection, and is not a valid argument against the reliability of the method when it is used for properly chosen samples.

References

[1] Henry M. Morris, Ed., *Scientific Creationism.* (San Diego: Creation-Life Publishers, 1974), 146-8.

[2] C. S. Noble and J. J. Naughton, "Deep-ocean basalts: inert gas content and uncertainties in age dating," *Science* 162 (1968), 265-6.

[3] G.B. Dalrymple and J. G. Moore, "Argon-40: excess in submarine pillow basalts from Kilauea Volcano, Hawaii," *Science* 161 (1968), 1132-5.

[4] J.G. Funkhouser and J. J. Naughton, "Radiogenic helium and argon in ultramafic inclusions from Hawaii," *Journal of Geophysical Research*, 73, No. 14 (1968), 4601-7.

[5] Andrew Snelling, "Radioactive dating failure," *Creation* 22 (1999), 18-21, reprinted on the website www.answersingenesis.org.

[6] Ian McDougall, H.A. Polach and J.J. Stiff. 1969. "Excess radiogenic argon in young subaerial basalts from the Auckland volcanic field, New Zealand." *Geochimica et Cosmochimica Acta*, 33 1969, 1485-1520.

[7] Morris, *Creationism*, 148.

[8] Snelling, "dating failure," 21.

25 Ages of Rocks in the Grand Canyon of Arizona

The Grand Canyon of Arizona provides an awesome and inspiring vista of rock layers of varying colors and thicknesses. A peek over the edge of the South Rim in the vicinity of Grand Canyon Village reveals a canyon about 5000 feet deep, and about 10 miles wide from rim to rim at that location. The vista is so large and so awe-inspiring that it prompted a friend to comment, "Every time I lifted the camera to my eye to take a picture, I was impressed most by how much I was leaving out!"

Designated a National Park in 1919, the Canyon attracts nearly five million visitors per year. The visitors come from all over the world, and include many geologists who come to see and study the rocks. The rocks are of many types: metamorphic schist with igneous intrusions in the lower part of the Canyon, layered sedimentary rocks above. The sedimentary rocks include sandstones deposited in a marine (ocean water) environment and containing fossils of marine creatures, other sandstones deposited on a land surface containing fossils of land plants, and some sandstone with evidence of having been deposited as wind-blown dunes, some shale layers deposited in ocean water with marine fossils, other shale deposited on land surface with plant fossils, and limestone deposited in ocean water with a large variety of marine fossils. These layered rocks bear testimony to processes of deposition in changing environments, from marine to land and back to marine, with periods of surface erosion between several of the sedimentary formations, leaving sharp contacts between the new sequence of deposition and the eroded surface of the layer below.

The canyon walls consist of a huge complex of cliffs and canyons that expose a very thick sequence of rock deposits to view, and, with a bit of difficulty, to close-up study. First investigated in detail by exploring parties led by John Wesley Powell in 1869 and 1871, the Canyon has long been considered by geologists to be a magnificent "window into the past." Fossil correlations identify the nearly horizontal sedimentary layers in the upper part of the Canyon as belonging to the Paleozoic Era, extending from the Tapeats Sandstone of Cambrian age at the rim of the Inner Gorge near Grand Canyon Village to the Kaibab Limestone of Permian age at the South and North rims of the Grand Canyon. Below the Tapeats Sandstone, and therefore older, are tilted sedimentary layers known as the Grand Canyon Supergroup; these are well exposed in the lower part of the Canyon east of Grand Canyon Village. Below the Supergroup, and therefore older still, are metamorphic rocks that have been intruded by igneous bodies of rock; these are exposed in the lowest part of the Canyon from east of Clear Creek, a few miles upstream from the bridges crossing the Colorado River on the South Kaibab and Bright Angel hiking trails, and extending downstream for a distance of about 160 miles. The steep cliffs of the Inner Gorge, visible from Grand Canyon Village, display these metamorphic rocks with igneous intrusions immediately below the layered Tapeats Sandstone.

That sequence of various rock types, deposited under various climatic conditions, led geologists to the conclusion that those rocks represented a considerable span of time in Earth's history.

Ages of the rocks

When it became possible to measure the ages of rocks based on radioactivity, those methods were applied to the rocks of the Grand Canyon, of course. The results of those measurements, using various isotope pairs, range up to 1.7 billion years.

Those results are unacceptable to those who promote the view that Earth is a recent creation, and various publications representing that view have lodged arguments against the validity of those results. Probably the most detailed of those arguments are presented in the publication *Grand Canyon: Monument to Catastrophe.*[1] Chapter 6 of that document is entitled "Are Grand Canyon Rocks One Billion Years Old?"

We will present some evaluation of the discussion in *Monument to Catastrophe* here in simple prose, and the technical details will be presented in Appendix F for those who wish to examine those details.

The argument(s)

The discussion in *Monument to Catastrophe* centers on the age measurement method that uses the radioactive isotope rubidium-87, which decays with a half-life of 49 billion years to form strontium-87. The very long half-life of Rb-87 means that only a small amount of Sr-87 is produced by Rb-87 decay, even for very old rocks. Consequently, the method works OK only for minerals that contain an appreciable amount of rubidium. A graphical method using the ratios of these two isotopes to Sr-86, called the "isochron" method, has been developed to measure ages in rocks that do not contain rubidium-rich minerals (discussed in more detail in Appendix F).

The professional scientific literature includes a report in which the isochron method of Rb-87/Sr-87 was applied to the Cardenas Basalts, several volcanic lava flows that are found within the sedimentary rocks of the Grand Canyon Supergroup, the tilted layers that lie below the horizontal Tapeats Sandstone. The age of the Cardenas Basalts, as measured by the Rb-87/Sr-87 method, was reported to be 1.09 billion years.[2]

There are also some volcanic cones and lava flows on the upper surface on the Uinkaret Plateau north of the western part of the Grand Canyon. These lie on top of the Kaibab Limestone, the uppermost sedimentary layer found in the Grand Canyon. Some of the lava flowed into the Grand Canyon, mantling the north wall of the Canyon and blocking the flow of the Colorado River for a time, until the river eroded its way through the volcanic deposit. Those lava flows are obviously recent, lying on top of the sedimentary layers of the Grand Canyon, and therefore younger than those sedimentary rocks. The author of *Monument to Catastrophe* collected some samples of lava from the Uinkaret Plateau and arranged to have them analyzed in commercial laboratories for rubidium and strontium. Then the ratio of Sr-87/Sr-86 was plotted vs. Rb-87/Sr-86 on an "isochron" diagram. Although all the samples contain very small amounts of rubidium - ideally, there would be a greater spread of rubidium values - the points lie along a very nice straight line. The "age" of these recent lava samples, determined by the Rb-87/Sr-87 isochron method, was found to be 1.34 billion years.[3] That result can't be correct; these are recent lavas, and yet the isochron age is greater than that of the Cardenas Basalts, which lie far lower in the sequence of Grand Canyon rocks.

The conclusion in *Monument to Catastrophe* is that radiometric age measurements by the Rb-87/Sr-87 isochron method are unreliable. With some additional comments, the conclusion of unreliability is extended to <u>all</u> radiometric measurements of Grand Canyon rocks, with the concluding rhetorical question:

"Has any Grand Canyon rock been successfully dated?"[4]

The rest of the story

The conclusion that the Rb-87/Sr-87 method is unreliable for measuring ages of volcanic rocks was already known and published in the professional scientific literature before the publication of *Monument to Catastrophe*.

The problem is not with the logic of the method, but with the samples; basaltic volcanic rocks do not provide samples that meet important conditions for the Rb-87/Sr-87 method to be valid. Several problems encountered in using the Rb-87/Sr-87 methods for volcanic rocks are summarized in Dickin in both the first edition of his book in 1995 and the second edition in 2005; in the second edition he began his discussion of dating igneous rocks by the Rb-87/Sr-87 method with the statement:

"The Rb-Sr method has largely been superseded as a means for dating igneous rocks."[5]

Conditions required for the Rb-87/Sr-87 isochron method to be valid for measuring the ages of rocks (the time of eruption for volcanic rocks) include the following:

Condition #1: The ratio of the element rubidium to the element strontium in the molten source of the volcanic eruption must be uniform throughout, and the ratio of the isotope Rb-87 to Sr-86 and the ratio of the isotope Sr-87 to Sr-86 must be uniform throughout the molten magma.

Dickin states that this condition is not generally met for basaltic volcanic rocks.[6]

To provide evidence supporting this statement, he refers to a report on studies of volcanic basaltic rocks in islands of the southwest Pacific Ocean. The "isochron age" of samples from those rocks obviously did not provide the time of the eruption of these recent volcanic rocks.[7]

Condition #2: There must be no separation of the element rubidium from the element strontium, nor for any of the isotopes involved from the other isotopes by chemical/mineralization processes (called "fractionation") taking place within the molten magma prior to the eruption of the volcanic lava.

Dickin states that this condition is not generally met for continental igneous rocks.[8]

To provide evidence supporting this statement, he refers to a report on studies of Rb-87/Sr-87 in continental igneous rocks, including both volcanic and plutonic rocks.[9]

Condition #3: The molten magma must not incorporate materials from, nor react chemically with, the rocks through which the molten magma passes from the magma chamber to the surface.

Dickin states that this condition is not generally met for basaltic volcanic rocks.[10]

To provide evidence supporting this statement, he refers to a publication which reported finding materials in lava samples from Scandinavia that had been derived from rocks through which the lava had passed during eruption to the surface.[11]

More details regarding those failures of volcanic samples to meet the necessary conditions for valid dating by the Rb-87/Sr-87 isochron method are found in Appendix F.

The reports referred to by Dickin were published a couple of decades before the publication of *Monument to Catastrophe*. So it was no surprise to find that the Rb-87/Sr-87 isochron for the samples from the Unikaret Plateau reported in *Monument to Catastrophe* provided an obviously incorrect age for those lava flows.

Other age measurements on Grand Canyon rocks

In addition to the Rb-87/Sr-87 measurements discussed above, radiometric ages have been determined for various samples of Grand Canyon rocks using various isotope pairs. The reports listed below are taken from a 1986 publication that lists the results of radiometric dating of 1,688 samples of rocks in Arizona, including 25 samples from the Grand Canyon.[12] That publication includes references to the original papers reporting those measurements, and we will not include references to those original papers here. We will not include any of the Rb-87/Sr-87 measurements.

The oldest rocks in the Grand Canyon, those lowermost in the sequence throughout the length of the canyon, include metamorphic rocks - formed from previously existing rocks by heat and pressure - and igneous rocks - formed as intrusions into those metamorphic rocks or deposited along with the metamorphic events. These rocks contain minerals that generally meet conditions for radiometric age measurement. The results for samples of the metamorphic rock probably determine the time at which metamorphism occurred, rather than the time when the original rocks were deposited, because the isotopes used for age measurement are likely to migrate during the metamorphic process, "resetting" the isotopic clocks.

Two measurements have been made on these older rocks using the U-238/Pb-206 method. One is a sample of Zoroaster granite, an igneous rock, and one is a sample of Zoroaster gneiss, a metamorphosed rock associated with the granite and likely formed at about the same time as the granite. The granite has an age of 1,705 million (1.7 billion) years, and the gneiss an age of 1,675 million years.

Three measurements of the age of the older metamorphic rock have been made by the K-40/Ar-40 method. The results are: 1,415 million years, 1,349 million years, and 1,255 million years. These results are somewhat less than the U-238/Pb-206 ages of the granitic rocks associated with the metamorphism. The younger age determined for the metamorphic rocks may be the result of some loss of argon during an extended span of elevated temperature following the processes of metamorphism.

Five measurements of the age of samples from the Cardenas Basalt lava flows by the K-40/Ar-40 method have been reported in the professional literature, in addition to the measurements by the unreliable Rb-87/Sr-87 method considered earlier. The results are: 853 million years, 843 million years, 820 million years, 801 million years, and 790 million years.

The phenomenon of "excess" argon-40 in volcanic rocks (see Chapter 24) was not considered in those measurements, so those ages may be a bit higher than the correct ages. According to reports in the professional scientific literature, the amount of error introduced into the age of recent lava samples by excess Ar-40 has ranged from very little up to a few million years, although it might be as much as a few tens of millions of years in rare cases. It certainly introduces a large potential error in samples that are only a few million years old or younger, but the potential error is a progressively smaller percentage of the actual age as the age of the sample increases. That being so, the true age of the Cardenas Basalts is likely to be greater than 700 million years, though probably less than the results reported above.

It is not entirely clear to this author that the problem of excess Ar-40 could be avoided by doing a graphical plot of K-40/Ar-40 measurements, extrapolating the line to zero potassium to determine the amount of excess argon, as was performed in *Monument to Catastrophe*.[13] The isochron method would assume that the amount of excess Ar-40 was the same in all samples, which may well not be the case. The result of 715 million years produced by that procedure, however, may well be close to the correct age.

There are also some reported ages by the K-40/Ar-40 method and by the Ar-40/Ar-39 method (a modification of the K-40/Ar-40 method that is designed to avoid the problem of excess argon) for the igneous dikes and sills associated with the Cardenas Basalts that were intruded into sedimentary rock, and cooled and crystallized below the surface. These were probably derived from the same magma source as the Cardenas Basalts, which did reach the surface as lava flows. Known as "diabase (a type of rock) dikes and sills" associated with the Cardenas Basalt lava flows, the ages of two samples by the K-40/Ar-40 method are 954 million years and 913 million years; the ages of two samples by the Ar-40/Ar-39 method are 907 million years and 904 million years.

In addition to these ages of older Grand Canyon rocks, there have been some measurements by the K-40/Ar-40 method of the ages of the lava flows on the Unikaret Plateau, the lava flows from which the author of *Monument to Catastrophe* obtained samples for Rb-87/Sr-87 analysis. Six samples in the list published by the Arizona Bureau of Geology range from 0.01 million (=10,000) to 3.67 million years. Those K-40/Ar-40 ages are suspect because of the possibility of excess argon in volcanic rocks.

However, the ages of samples from the Uinkaret Plateau have also been measured by other methods. The ages of those samples have been determined by Ar-40/Ar-39, a method which is designed to avoid the excess Ar-40 problem. Measurements were also done with a more recently developed method, called "cosmogenic age measurement," that measures the time during which the deposit has been at Earth's surface, where it has been exposed to the neutrons resulting from bombardment of Earth's atmosphere by cosmic rays from outer space (as described in Chapter 21). The results of the Ar-40/Ar-39 and of the cosmogenic age measurements are in good agreement. According to these age measurements, the lava flows that actually entered the canyon to form temporary dams in the Colorado River were extruded over the period of time ranging from the oldest at 600,000 years ago to the youngest at 100,000 years ago.[14] So the radiometric ages of these lava flows by methods other than rubidium-strontium are in agreement with the observation that they are "recent" events in the history of the Grand Canyon.

If truth be known, we can answer the question, "Has any Grand Canyon rock been successfully dated?" with a resounding "YES!"

References

[1] Steven A. Austin, Ed., "Are Grand Canyon Rocks One Billion Years Old?" in *Grand Canyon: Monument to Catastrophe*, (Santee, CA: Institute for Creation Research, 1994), 111-32.

[2] Edwin H. McKee and Donald C. Noble, "Age of the Cardenas Lavas, Grand Canyon, Arizona," *Geological Society of America Bulletin*, 87 (1976), 1188-90.

[3] Austin, "Grand Canyon," 124-5.

[4] Austin, "Grand Canyon," 129

[5] Alan P. Dickin, *Radiogenic Isotope Geology*. (Cambridge, UK: Cambridge University Press, 2005), 43.

[6] Dickin, *Radiogenic*, 45-46.

[7] S.S. Sun and G.N. Hanson, "Evolution of the mantle: geochemical evidence from alkali basalt," *Geology* 3 (1975), 297-302.

[8] Dickin, *Radiogenic*, 46.

[9] C. Brooks, D.E. James, and S.R. Hart, "Ancient lithosphere: its role in young continental volcanism," *Science* 193 (1976), 1086-94.

[10] Dickin, *Radiogenic*, 46.

[11] R.D. Beckinsale, R.J. Pankhurst, R.R. Skelhorn, and J.N. Walsh, 1978. "Geochemistry and petrogenesis of the early Tertiary lava pile of the Isle of Mull, Scotland." *Contributions to Mineralogy and*

Petrology 66 (1978), 415-27.

[12]S.J. Reynolds, F.P. Florence, J.W. Welty, M.S. Roddy, D.A. Currier, A.V. Anderson, and S.B. Keith, *Compilation of Radiometric Age Determinations in Arizona*, Bulletin 197 of the Arizona Bureau of Geology and Mineral Technology, Geological Survey Branch, 1986.

[13]Austin, "Grand Canyon," 122.

[14]C.R. Fenton, R. J. Poreda, B.P. Nash, R.H. Webb, and T.E. Cerling, "Geochemical Discrimination of Five Pleistocene Lava-Dam Outburst-Flood Deposits, Western Grand Canyon, Arizona." *Journal of Geology*, 112 (2004), 91.

26 Errors and Corrections

Stephen Jay Gould wrote a column entitled "This View of Life" for the monthly journal *Natural History* over a span of 28 years until his death in 2002, and some collections of his columns have been published in book form. Although I certainly do not share Gould's agnosticism, I have found some valuable insights in some of his columns.

Several years ago, just after Judge Overton had ruled that teaching "creation science" in Arkansas public schools would violate the separation of church and state, Gould wrote an article commenting on that case.[1] You may remember that Judge Overton based his opinion in part on the judgment that "creation science" is not science, but is a sectarian teaching based on the Christian Scriptures. Gould agreed with Judge Overton's ruling and with his judgment that "creation science," a.k.a. "scientific creationism," is not science.

I think that Judge Overton was partially mistaken in his judgment that "scientific creationism" is not science. He <u>was</u> correct in concluding that the brand of "scientific creationism" promoted by the Institute for Creation Research is ultimately based on their interpretation of the Christian Scriptures rather than on evidence from the study of the physical world, and, to that extent, it is not science. The last chapter in the book *Scientific Creationism* plainly states the claim that the Bible is the ultimate authority on scientific matters.[2]

But the publications that promote the view that Earth is a recent creation <u>do</u> publish some claims about matters that can be investigated by the methods of science, such as the ages of rocks, the Second Law of Thermodynamics, and others that have been considered in the earlier chapters of this book. There are a few more such claims that have been evaluated in the book *Science Held Hostage*.[3]

In the judgment of this author, the teaching of "scientific creationism" should be excluded from instruction in public schools in Arkansas and elsewhere, but not because the publications that promote the view that Earth is a recent creation don't contain some scientific claims. It should be excluded because the scientific claims published as "scientific creationism" or "creation science" <u>can be</u> and <u>have been</u> investigated, and have been found to be mistaken.

In another of Gould's columns he wrote about the "Nebraska Man" case, discussed also in Chapter 12 of this book. Gould concluded that the "Nebraska Man" case demonstrates the strength of science, rather than weakness, since it demonstrates the admission and correction of error. In that column Gould challenged the "scientific creationists" to emulate the scientists involved in the "Nebraska Man" episode, and to openly admit and correct errors that have been published in the books, articles, and websites that promote the view that Earth is a recent creation. Gould made special reference to the Paluxy River tracks, which had been claimed at an earlier time as evidence that dinosaurs and humans are contemporaries. Gould wrote:

"The real message of *Hesperopithecus* [Nebraska Man] proclaims that science moves forward by admitting and correcting its errors. If creationists really wanted to ape the procedures of science, they would take this theme to heart. They would hold up their most ballyhooed, and now most thoroughly discredited, empirical claim - the coexistence of dinosaur and human footprints in the Paluxy Creek beds near Dallas - and publicly announce their error and its welcome correction."[4]

We have discussed how the scientific community deals with deliberate deceit and fraud in science (Ch. 11, the Piltdown hoax). We have discussed how the scientific community deals with errors in science (Ch. 12, Nebraska Man). Let us now consider how the publications that promote the view that Earth is a recent creation treat recognized error.

Paluxy River tracks

The claim was made in publications that promote the view that Earth is a recent creation that human tracks were found along with dinosaur tracks in the Glen Rose (limestone) formation, exposed in the bed of the Paluxy River near Glen Rose, Texas. The conclusion published with that claim is that humans and dinosaurs must have been contemporaries. This claim was published in *Scientific Creationism* in the following words:

> "One of the most spectacular examples of anomalous fossils is the now well-known case of the Paluxy River footprints, in the Cretaceous Glen Rose formation of central Texas. Here, in the limestone beds are found large numbers of both dinosaur and human footprints."[5]

That claim was elaborated in the book *Tracking Those Incredible Dinosaurs and the People Who Knew Them.*[6]

The tracks from the Paluxy River had come to the attention of Roland T. Bird, a paleontologist collecting fossils for the American Museum of Natural History, in 1938. Spending a night in Gallup, New Mexico on his return from fossil hunting in western states, he was told about some fantastic "human" tracks that were displayed in a store window in a small town near Glen Rose, Texas. He traveled to that town, and found what appeared to be near perfect replicas of giant human footprints, but more careful examination showed them to be "clearly sculpted copies of the bottom of an imaginary foot."[7] He was also told of some dinosaur footprints in the store owner's other store in a nearby town, but he reported that "they appeared to be prints of a medium-sized carnivore, but there was something too suspiciously perfect" about the prints.[8] Those tracks also turned out to be fake, apparently carved by the same artist as the giant "human" tracks, and sold to the store owner when he had visited Glen Rose, Texas. Next day R.T. Bird drove to Glen Rose where he was able to contact some local people who knew about genuine dinosaur tracks in the bed of the Paluxy River nearby. He found a remarkable set of tracks of a large sauropod with tracks of a good-sized carnivore alongside, and returned a year and a half later, with a crew of workmen, and excavated a trackway that is now part of the Apatosaur (=Brontosaur) display at the American Museum in New York City. (But there is not another word in his book about human footprints, real or fake.)

In 1986 an article was published that reported a detailed study of the supposed human tracks, citing evidence that they are, in fact, not made by humans but by an unidentified bipedal dinosaur.[9] Soon thereafter a team of investigators from the Institute for Creation Research visited the site and did extensive observations. Their conclusion, also, was that the tracks very likely were made by a bipedal dinosaur and not by humans. The announcement stated that "it would now be improper for creationists to continue to use the Paluxy data as evidence against evolution."[10] This admission of error demonstrates that "scientific creationists" <u>are</u> able and willing to admit errors.

In 1987, soon after this admission of mistaken identification, the book *The Great Dinosaur Mystery and the Bible* was published by the Institute for Creation Research. The central theme of the book is that dinosaurs and humans were contemporaries, but, in accordance with the just-announced admission of error, there was no mention of the Paluxy River tracks.[11]

But there's something interesting about that admission. The book *Scientific Creationism* was published in 1974 by the publishing arm of the Institute for Creation Research. The back cover says that this book is "the most practical, well-organized resource handbook on scientific creationism available today." The claim that the Paluxy River footprints demonstrate that humans and dinosaurs were contemporaries appears on pages 122-123 of that edition, first printing. In 1985 the 12th printing of *Scientific Creationism* was published with a banner across the front cover stating "UPDATED AND ENLARGED." The back cover of this 12th printing says that this book is "the most practical, well-organized resource handbook on scientific creationism available today," in the same words as the back cover of the 1974 edition. The argument that human and dinosaur footprints are found together in the Paluxy River site appears in this 12th printing, still on pages 122-123, mostly unchanged from the 1974 first printing, with only a replacement of one sentence to include the publication *Tracking Those Incredible Dinosaurs and the People Who Knew Them* as further support for that claim, and an added footnote reference to that book.

The 24th printing of *Scientific Creationism* was being sold by Christian bookstores and by the publisher as recently as the year 2009. The banner about "updated and enlarged" is not on the front cover, but the statement that this was the "updated and enlarged" version is stated on the back cover. The argument that human and dinosaur tracks are found together at the Paluxy River site appears in this 24th printing, still on pages 122-123, unchanged from the 1985 printing, although the admission that this claim is mistaken had been published in 1986. These are not simply old copies that were published before the admission of error, and that continue to sit on the shelf. These are newly published copies that continue to publish the claim that was supposedly admitted as an error by representatives of the publisher at an earlier time.

Additionally, the website www.creationevidence.org continues to promote the claim that human tracks are found with dinosaur tracks at the Paluxy River site, and that humans and dinosaurs lived together on Earth.

Isn't that interesting?

Dust on the moon

Another of those publicly admitted errors in publications that promote the view that Earth is a recent creation has to do with the accumulation of meteoritic dust on the moon.[12] Some of the "scientific creationists" did their own study of that matter, and published the conclusion that the data actually support an old age for the moon.[13] These findings contradict the earlier claim in *Scientific Creationism* that the dust on the moon supported a recent origin of the moon. This admission of error demonstrates that "scientific creationists" <u>are</u> able and willing to admit errors.

However, this case of admission of error regarding the dust on the moon is remarkably similar to admission of error in the case of the Paluxy footprints, discussed above. The argument that the-moon-must-be-young-because-of-the-dust appeared on page 152 of the 1974 edition of *Scientific Creationism* (at which time it had already been conclusively refuted by unmanned spacecraft and manned Apollo visits to the moon), and continues to be published in the most recent printings of *Scientific Creationism*, still on page 152, as in the 1974 edition.

Isn't that interesting?

Further comments

The publications that promote the view that Earth is a recent creation pooh-poohed the correction of error by the scientists involved in the "Nebraska man" case as being "begrudgingly concluded," and claiming that "the truth leaked out slowly and obscurely," and that the "facts were not considered generally newsworthy," although the correction was published in the same professional journal as the erroneous identification had been published, and by the same scientists as had been involved in the erroneous identification.

The publications that promote the view that Earth is a recent creation pooh-poohed the exposure and correction of the "Piltdown Man" fraud by the scientific community as follows:

"Evolutionary scientists would rather forget about it, or turn it into a cute detective story. ... In fact, historical revisionists claim that the Piltdown hoax is evidence of the *strength* of evolutionary 'science' because it is self-correcting - given time, scientists will weed out mistakes and errors. ... But the real question is: why did it take so long for scientists to discover such an obvious fraud? The obvious answer is that scientists are so blinded by their faith in human evolution and 'missing links' that they 'see' things that just aren't there."[14]

So, in what sorts of publications has the admission and correction of error been published by the organizations that promote the view that Earth is a recent creation? 1) In *Acts and Facts*, a monthly newsletter published by the Institute for Creation Research, formerly El Cajon, CA, now Dallas, TX. 2) In *Creation ex nihilo*, published by Creation Ministries International, Queensland, Australia. 3) On the website www.answersingenesis.com and on the website for Creation Ministries International. How wide a readership do you suppose those publications and websites enjoy among the general public, and in the scientific community?

And what effect has the admission and correction of error had on later publications of organizations that promote the view that Earth is a recent creation? In the case of the book *Scientific Creationism*, obviously, none.

Expectations for the future

To its credit, the website www.answersingenesis.org has published a list of fallacious claims under the title "Arguments that should never be used" in defense of the view that Earth is a recent creation. (A similar list is found on the website of Creation Ministries International.) All of the entries on the list are claims that have been made in the past by proponents of the view that Earth is a recent creation in various publications, either claiming support for the view that Earth is young, or claiming evidence that refutes the "old-Earth" position that has been accepted by most scientists as valid. Some of those fallacious claims are characterized as being "Arguments that should never be used," and others as "Arguments that should be avoided (because further research is still needed, … .)"[15]

The list of arguments that should not be used includes the "dust on the moon" argument, and the "Paluxy river footprints" described above; while these are rejected on the website www.answersingenesis.org, they still appear in various "young-Earth" publications, including the edition of *Scientific Creationism* being sold in bookstores as recently as 2009.

It is likely that the www.answersingenesis.org list of formerly-used arguments that should no longer be used to claim scientific support for the view that Earth is a recent creation will grow longer year by year, as one after another current claim is found to be mistaken or misleading. Most, perhaps all, of the arguments currently found on www.answersingenesis.org claiming scientific support for the view that Earth is a recent creation will likely fall by the wayside as time goes by and truth becomes known. At what point, do you suppose, will we become convinced that <u>none</u> of these claims of scientific support for the view that Earth is a recent creation is valid? At what point, do you suppose, will we become convinced that <u>none</u> of the claims of scientific support for the "flood geology" explanations for the rocks in the Earth's crust is valid?

If we suppose that there were, say, 200 published claims of scientific support for the view that Earth is a recent creation, and 199 of them have been investigated and have been found to be mistaken, would anyone still cling to the 200th claim, the only one not discredited as yet, and persist in the claim that there is scientific support for the view that Earth is a recent creation on the basis of that one claim?

References

[1]Stephen J. Gould 1982. "Genesis vs. Geology," *The Atlantic Monthly* 250, No. 3 (September 1982), 10.

[2]Henry M. Morris, Ed., *Scientific Creationism*, (San Diego: Creation-Life Publishers, 1974), 203, 255.

[3]Howard Van Till, Clarence Menninga, and Davis A. Young, *Science Held Hostage*. (Downers Grove, IL: Intervarsity Press, 1988).

[4]Stephen J. Gould, "An Essay on a Pig Roast," *Natural History* 98

(April 1989), 15. Reprinted in *Bully for Brontosaurus*, (New York: W.W. Norton and Company, 1991), 447.

[5]Morris, *Creationism*, 122.

[6]John D. Morris, 1980. *Tracking those Incredible Dinosaurs … and the people who knew them.* (San Diego: Creation-Life Publishers, 1980).

[7]Roland T. Bird 1985 *Bones for Barnum Brown*, (Fort Worth: Texas Christian University Press, 1985), 146.

[8]Bird, *Bones*, 147.

[9]Glen Kuban, "The Taylor Site 'man tracks,'" *Origins Research* 9, (Spring/Summer 1986), 7.

[10]John D. Morris, "The Paluxy River Mystery," *Acts and Facts* (January 1986).

[11]Paul S. Taylor *The Great Dinosaur Mystery and the Bible.* (San Diego: Master Books Publishers, 1987).

[12]Van Till, *Hostage*, 67-82.

[13]Andrew A. Snelling and David E. Rush, "Moon Dust and the Age of the Solar System," *Creatio ex Nihilo Technical Journal*, 7, No.1(April 1993), 2-42. Also available by link on www.answersingenesis.org.

[14]Michael Matthews, "A century of fraud" in www.answersingenesis.org/docs2003.

[15]www.answersingenesis.org/get-answers/topic/arguments-we-dont-use, 11/30/2009.

27 Unsolved Puzzles

There are a good many things about the natural world that have not been satisfactorily explained by the methods of science. There are still a great many topics available for Ph.D. thesis study. But the realization that the scientific study of God's world has not yet solved all the puzzles doesn't mean that the results of scientific study are totally unreliable. A great many aspects of our natural world that puzzled earlier generations have been satisfactorily explained.

As was pointed out in Chapter 1, scientific methods are unable to provide answers to questions of purpose and meaning. We're not talking about those questions. We need to seek elsewhere for answers to those questions.

For answers to questions and puzzles about our physical world, however, the methods of science have been remarkably successful at finding explanations. The explanations are not perfect, and every scientific law, theory, and hypothesis is subject to continued evaluation and possible modification. That's what is meant when we say that the results of scientific study are "tentative." If truth be known, the perspective that scientific explanations are always open to evaluation and possible modification is one of the greatest strengths of the scientific methods.

Many of the results of the scientific study of God's world have been surprisingly different from what anyone had expected. Two examples: 1) Before the time of Galileo, everyone conceived of the moon as a "celestial" object, totally different from the "terrestrial" matter and characteristics of Earth. When Galileo pointed his telescope at the moon, however, he found mountains that cast shadows in the sunlight, valleys and ridges, all entirely Earth-like. He also observed, for the first time, that the planet Venus displays phases ranging from crescent to full, just as the moon does, and confirming that Venus, at least, revolves around the sun, implying that other planets, including Earth, also revolve around the sun. All very new and surprising results. 2) Before the launch of the first spacecraft, Sputnik, in 1957, everyone acquainted with magnetism and electricity and electrons had expected to find some radiation in space. But no one had predicted or suggested that a very high concentration of electrons would be found in the region of Earth's magnetic equator, now known as the "Van Allen radiation belts." In retrospect, that phenomenon could/should have been expected, but the discovery was very surprising to even the best-informed scientists.

Who can predict what surprising discoveries are yet in store for us? We know how to ask some of the questions, but we do not yet know how to find the answers to those questions. Sometimes it takes a long time to find answers to our questions. Again, two examples: 1) In 1687 Sir Isaac Newton suggested a universal gravitational force that holds the moon in its orbit around the Earth, and holds the planets in their orbits around the sun. He was challenged by some of the doubters to "demonstrate" the existence of such a force by a laboratory experiment, but he was unable to do so. It took a hundred years before Cavendish succeeded in doing such an experiment, thus enabling us to "weigh the Earth." 2) In the mid-1800's the studies in thermodynamics led to the conclusion that the sun's energy could not possibly be from combustion of fuel by combining with oxygen, as occurs in fires on Earth, because the sun would long ago have consumed all of its fuel. The puzzle of how the sun produced its energy was finally solved in the 1930's with the suggestion that the nuclei of hydrogen atoms undergo a "fusion" reaction at high temperatures, forming helium and releasing prodigious amounts of energy. That is the process that was put to use (misuse?) by humans in developing a "hydrogen bomb." That process is now being studied in several laboratories in the hope of finding a way to control the reaction for the production of electrical energy.

Many more examples could be submitted from the history of

science.

As new discoveries are made, and old puzzles solved, our concepts of our physical world change. By contrast with people of long ago, we now believe that the Earth is spherical in shape, the planets revolve around the sun, matter is made up of small, sub-microscopic particles (atoms), which in turn consist of still smaller particles, etc. etc. How exciting! Who can imagine what yet lies in store for us?

The publications that promote the view that Earth is a recent creation are very fond of pointing out the fact that scientists have sometimes changed their (and our) concepts of what the universe around us is like. The implication, and sometimes explicit accusation, is "How can you believe <u>anything</u> those scientists tell us? They are always changing their minds about this or that! They can't even explain 'such and such;' when are they finally going to tell us how things really are?"

If truth be known, the willingness to change our concept of the universe in the face of new discoveries and modified understanding is one of the great strengths of science. If we never changed our concepts of how the universe works, we would still be living in superstitious fear of every sun eclipse, and subject to devastation by every new contagious disease.

If truth be recognized, there is much yet for us to learn about God's world. The scientific enterprise is one avenue by which to make some progress in the direction of understanding. True enough, new discoveries challenge some of our long-standing traditional ways of looking at or thinking about some things. Would our understanding of God's world be improved if we rejected the reasons for believing that the Earth rotates and revolves around the sun? Would it be better for us to revert to the time when everyone was convinced that the Earth is stationary in space, as some Christians advocated in opposition to Galileo? Should all of us Christians join the Flat Earth Society? (Yes, there is such an organization.)

I don't think that any of us would choose to reverse our understanding of Earth's shape and motions in the solar system.

Have some of our scientific discoveries been used for evil purposes, rather than for good? Sad to say, it is true. The problem is in the attitudes of the human heart, however, and cannot properly be ascribed to the understanding of God's world.

Unsolved puzzles demonstrate that our human understanding of God's world is incomplete, and changing concepts in scientific explanation demonstrate that our human understanding of God's world is imperfect. Will we ever be perfect, and will our concepts of God's world ever be complete? Not likely in this life!

If truth be known, modern scientific investigation by modern scientific methods is far more trustworthy than the rejection of it.

Of course, rejecting the view that Earth is a recent creation, and rejecting the view that nearly all deposits of fossil-bearing rocks were deposited during the year-long universal flood described in Genesis 8-10, does not mean that you must reject the Christian faith. Christian faith does not depend on how old you think the world is, nor on how you think the rocks were set in place, nor or your answer to the question of whether or not the sun revolves around an Earth that is stationary in space. "If you confess with your mouth 'Jesus is Lord,' and believe in your heart that God raised him from the dead, you will be saved." Romans 10:9.

28 Reflections

We have finished almost thirty chapters dealing with published claims of scientific support for the conclusion that Earth is a recent creation, not more than several thousand years old. We have found that the claims we have considered are mistaken and/or grossly misleading. We have commented on some stories that are mistakenly construed to cast doubt on the validity of scientific evidence supporting the conclusion that Earth and the universe are billions of years old. So, what's next?

There are many more claims of scientific support for the thesis that Earth is a recent creation. It would take several hundred more pages to examine each one to determine whether or not the claim is valid. But we will not add more in this book; the claims we have considered here give you a pretty good idea of what's found there.

How many claims of scientific support for the view that Earth is a recent creation must be shown to be mistaken or grossly misleading before we say: "Enough! There **IS NO** scientific support for that viewpoint."

One may choose to accept and believe the view that Earth is a recent creation for a variety of reasons. He/she may understand Scripture to teach that view; she/he may have strong ties of allegiance to a Christian church community that holds to that view; etc., etc. [add your own reasons, if you wish] There is no dishonor in holding to such a view.

The preponderance of evidence from the scientific study of the universe that God created, however, supports the view that Earth and the universe are billions of years old. If you adopt the view that Earth is a recent creation, the conclusions that we have reached in our consideration of the topics of this book suggest that you must reconcile yourself to doing so without the support of scientific arguments.

Science and Christian Faith

So, must we give up our commitment to the Christian faith if we choose to accept the results of scientific study of God's world as valid understanding? Certainly not! Must we give up our conviction that the universe is the product of God's creative genius and power if we choose to accept conclusions from the scientific study of that universe as valid conclusions? Certainly not!

There are two aspects of that affirmation that deserve a bit more comment: 1) the compatibility of natural science with Christian faith and Bible interpretation, and 2) the informed judgment about whether a proposed scientific explanation for our observations is valid, or not valid.

The history of the interactions of natural science and Christian communities over the past few hundred years is instructive for us today. Numerous discoveries resulting from scientific investigation of God's world have brought some Christians into disagreement with other Christians and non-Christians about the proper understanding of God's world and God's word. The issue of the motions of the Earth in the solar system - the case of Galileo and his disagreement with his church - is a prominent example. That and similar cases, those that have been resolved to nearly everyone's satisfaction, should help us to identify which approaches to understanding are likely to be helpful in resolving apparent inconsistencies in current issues raised by scientific study. "Those who are unacquainted with history are destined to repeat history's mistakes."

The Christian faith has maintained its viability through those disagreements. Many of the early scientists, those who pioneered the approach to understanding that we call "modern science," were Christians. Some historians suggest that the Judeo-Christian perspective was an important foundation for developing the methods of modern scientific investigation and explanation. Today, also, there are thousands of professional scientists who are committed to Christian faith; some of them have achieved international prominence. We may be confident that the scientific discoveries of the present and the future will not destroy our belief in God and Jesus Christ. Science, properly conceived, neither denies nor proves the existence of God.

There are many books that promote the compatibility of Christian faith and natural science. Some of them are included in the list of books for further reading that follow this chapter. The basic premise of most of them is that God is made known to us through two major avenues; his revelation of himself to us in the Bible, and his revelation of himself as perceived through his handiwork in the physical universe as creator and governor of everything that exists. When both of those avenues of revelation are properly understood, there is no conflict or inconsistency between God's word and God's world; any inconsistency that arises is between science - our understanding of God's world, and theology - our understanding of God's word. So it is our understanding that is in need of evaluation and modification. Thus, scientists and theologians, hopefully working together, are resource people who can be helpful in resolving any perceived inconsistency in our understanding of God's world and our understanding of God's word.

A NOTE in passing: I had the experience, on one occasion, to carry on some correspondence with a fellow Christian who wrote to ask me some questions about science and theology. I was able to answer his questions only in part, and I suggested that he read two or three books that I thought would be useful to him in gaining some perspective in dealing with the issues he was asking about. He responded with a complaint: "I asked you some questions which you didn't answer, and all you do is want me to read some more books." My response to him, and my suggestion to you, dear reader, is that much wisdom and understanding has been recorded in books, and I don't know any method for gaining wisdom and understanding that doesn't include reading some books.

How do we know? (or How can we tell?)

Now, a few words about the more difficult question: "How do we determine whether any particular scientific claim is valid, or reliable?"

In Chapter 1 we noted that scientific explanations, or claims, are based ultimately on our observations of God's world. Those observations may be made directly with our senses, and also indirectly with the use of various instruments. To be valid, the scientific claim must be consistent with our observations. Any inconsistency with our observations calls for further explanation, or a modification of the scientific explanation, or, in case of persistent inconsistency, a replacement of the explanation with a better one.

That structure of scientific investigation and explanation has been hugely successful over the past few hundred years in solving puzzles that had not been solved before. There are also certain consequences, or characteristics, of that structure, that we should note:

1. A scientific explanation is always tentative. The phenomena being considered are always open to further investigation and observation, and the proposed explanation is always subject to possible modification.

a. It is rare indeed that later developments will permit a direct observation that confirms the validity of an earlier scientific explanation for some phenomenon or cluster of phenomena. One such case that comes to mind is the shape of the Earth; space travel has permitted us to "see" the spherical shape of Earth that was only deduced from indirect evidence at earlier times.

b. Even the widely accepted idea that Earth rotates daily and revolves around the sun annually is based on indirect evidence; there is no direct observation of those phenomena. Will there ever be? We don't know.

2. Whether a scientific claim (explanation) is valid or reliable is a matter of judgment. We ask questions like: 1) Is the evidence persuasive? 2) Are the observations themselves reliable? 3) Are there

known observations that are inconsistent with the claim? Are those observations reliable? 4) How many different kinds, or avenues, of evidence support that claim? 5) How does this claim fit with already established scientific understanding in closely related areas of study, and in the wider scientific understanding of the universe?

There are more questions, or more detailed ways to frame those questions, but that gives you the flavor.

3. So who is qualified to make judgments about the validity of a particular scientific claim? Well, of course, those who are well acquainted with the observations, the evidence, the wider context, any alternate explanations, etc.

So, like it or not, if you want to participate in making judgments about scientific explanations, you obligate yourself to become well acquainted with the observations, the evidence, the wider context, any alternate explanations, etc. Lacking that, your judgment lacks credibility.

4. What is meant by calling a scientific explanation "valid" or "reliable?" It means that the explanation is consistent with most or all of the evidence available to us at the present time, and that few or none of our observations are inconsistent with that explanation.

a. I think that we should talk about the validity of scientific explanations in terms of "level of confidence," rather than talking about "prove" or "disprove." A scientific explanation generally progresses from its first proposal through a sequence of testing until it becomes accepted with a high level of confidence by those qualified to judge, even while it always remains open to further evaluation.

b. The process of evaluating scientific explanations takes time. New or novel explanations are generally held at a lower level of confidence than those that have undergone extensive evaluation already and have survived the tests.

5. So where does that leave the person who has little or no scientific training, the non-science people who occupy the pews in our churches from Sunday to Sunday?

My suggestion to you, if you are interested in issues in science and Christian faith, is to place your confidence in people who are trained in science, people whom you know well enough to consider them trustworthy. I would suggest that you consider the judgment of the wider scientific community, both Christians and non-Christians. Don't focus on one or a small group of scientists, but consider the judgments of a wide range. And remember, please, that Christian faith does not stand or fall on the basis of any scientific explanation. Romans 10:9: "If you believe in your heart the Jesus is Lord, and confess with your mouth that God has raised him from the dead, you will be saved."

The scientist

The professional scientist is obligated to the general public for helping people understand not only the accepted scientific explanation or claim, but also the evidence supporting that claim. He/she should be expected to answer the question: "How do we (you) know that? What is the evidence?" when she/he is asked to do so. Helping students find the answer to that question is (or should be) the core of science teaching.

NOTE: Sometimes a scientific explanation becomes so embedded into our everyday thinking that we no longer question its reliability, or the evidence supporting that idea. Few students entering college would doubt that the Earth rotates daily and revolves around the sun annually. When reminded that it doesn't look like that in our casual observations - it is the sun that rises and sets, after all - few of those students can report any of the evidence that supports the idea that the Earth rotates and revolves around the sun.

It is with some sadness that I have to tell you that some prominent scientists and too many science writers and reporters do a deplorable job of communicating with the public, although many do very well. Too often, prominent spokespersons for science overstate the level of confidence with which some scientific explanations are held, with the arrogant attitude that seems to say, "Trust me. I know!" instead of honestly presenting evidence supporting the scientific claim and recognizing any serious questions that remain unanswered.

Sometimes, even the terminology that is employed by prominent scientific spokespersons is mistaken and misleading. I have often heard on the TV and radio, and read in the newspapers and news magazines, that "the theory of biological evolution is fact." But theories - scientific explanations - never were and never will be "fact." Scientific explanations attempt to account for our observations, the facts that can be ascertained and established as evidence, but the explanation always remains just that, an attempted explanation. It would be more helpful to the public, and more modestly honest for the scientist him/herself to say, "this theory/explanation is held to be reliable at a high level of confidence." And then, when appropriate, to summarize the evidence on which that claim is based, hopefully in language that is understandable to everyone who cares to listen.

A statement that is almost invariably included in presenting the "scientific method" in our school textbooks is that "scientific explanations are tentative." OK, then let us scientists talk as if we know that.

Perhaps with even more sadness and chagrin, I have to tell you that even the textbooks often deny what they have just said about scientific explanations being tentative, and they go on to claim (incorrectly) that "natural laws are invariant." The most that we are ever justified in claiming regarding our scientific explanations, including those we call "natural laws," is that we consider them to be reliable at a high level of confidence.

Let scientists tell it "like it is."

The theologian

The professional theologian is obligated to tell the general public, especially those of us who sit in the pews of our churches from week to week, about our understanding of the Scriptures. In presenting an interpretation of a passage of Scripture, he/she is, or should be, expected to provide reasons for preferring that interpretation over other possible interpretations. And she/he should do this in awareness of the alternate explanations that have been, or could be, adopted by serious students of the Scriptures. When asked, he/she should be expected to provide reasons for considering a particular interpretation to be more reliable, more in consonance with the whole of Scripture, than the alternatives.

Like science, theology is a human activity. Like scientists, theologians are less than perfect, and their work is less than complete, in their interpretation of the object of their study. Therefore, like the explanations of the scientist, we could/should say that theological interpretations are "tentative," and always open to further investigation and evaluation.

Most of us who sit in the pews of our churches from week to week do not regularly read the professional journals of theology, just as most of us do not regularly read the professional journals of science. Our contact with theology is through our preachers in their sermons, through our teachers in Sunday School lessons, and in our contact with fellow believers in our small group Bible study. If you are seeking more detailed theological discussions, you will do well to become acquainted with the professional theological literature; if you are seeking more detailed scientific discussions, you will do well to become acquainted with the professional scientific literature.

There is a core set of beliefs in the Christian religion, probably best represented in the statement known as the "Apostles' Creed," and confessed by nearly every Christian over many centuries. Beyond that core set of beliefs, there is an amazing range of interpretations of Scripture, leading to the formation of the amazing range of denominations in the Christian church of today. Each of us, in becoming Christian, has been taught from the perspective of a particular tradition, and each of us, in becoming adult adherents of a particular denomination, has made a choice of convictions from among that wide range of interpretations. It is often difficult to sort out which parts of our personal set of beliefs are essential to the Christian faith, and which parts are disposable parts of the tradition within which we came to faith in Jesus Christ.

We may have come to the conclusion that we have some reservations about some of the tenets of Bible interpretation that characterize every denomination, but that we still prefer the denomination that we have chosen as our own.

The pewsitter

So what are we, the people who sit in the pews of our churches from week to week, to do? How can we choose a perspective that is likely to survive any apparent conflict between the published results of scientific study and what we had thought to be the teachings of the Bible?

1. Hold tight to the core teachings of Scripture, as expressed in the Apostle's Creed and other so-called "ecumenical" statements of faith, those adhered to by all Christian churches.

2. Hold other theological interpretations, those that we find to be different among churches that profess the core beliefs, as being tentative, allowing for such differences among fellow Christians.

3. Hold scientific explanations to be tentative, being aware that some are accepted by the wider scientific community as being held at a high level of confidence, while others, especially recent ones, are less certain.

4. Be patient, both with respect to science and with respect to theology. Allow some time for consideration and further study. Don't rush to judgment, one way or the other.

5. In present-day apparent conflicts between science and theology, adopt attitudes that would have held up well in earlier apparent conflicts. Some knowledge of history is valuable here; in speaking of his study of the stars, moon, and planets, Galileo wrote, "this study teaches us how the heavens go, not how to go to heaven."

6. Expect/demand scientists to be honest and forthright in acknowledging doubts and differences regarding uncertain interpretations of God's world. Expect/demand theologians, including preachers, to be honest and forthright in acknowledging doubts and differences regarding uncertain interpretations of God's word.

Finding peace

My personal preference is to adopt the perspectives taught by reformer John Calvin and early Christian theologian Augustine.

John Calvin was well aware of the results of the scientific study of God's world as they were understood in his day. In his *Institutes of the Christian Religion*, Chapter V, entitled "The Knowledge of God Conspicuous in the Creation and Continual Government of the world," Paragraph 2, Calvin wrote:

"In attestation of his [God's] wondrous wisdom, both the heavens and the earth present us with innumerable proofs, not only of those more recondite proofs which astronomy, medicine, and all the natural sciences, are designed to illustrate, but proofs which force themselves on the notice of the most illiterate peasant, who cannot open his eyes without beholding them. It is true, indeed, that those who are more or less intimately acquainted with those liberal studies are thereby assisted and enabled to obtain a deeper insight into the secret workings of divine wisdom. …. To investigate the motions of the heavenly bodies, to determine their positions, measure their distances, and ascertain their properties, demands skill, and a more careful examination; and where these are so employed, as the providence of God is thereby more fully unfolded, so it is reasonable to suppose that the mind takes a loftier flight, and obtains brighter views of his glory."[1]

So John Calvin had high respect for the scientific study of God's world, and considered scientific knowledge to be an advantage in perceiving God's hand in the creation and government of the universe.

Concerning astronomy in particular, Calvin wrote (commenting on Genesis1:16 and the results of the scientific conclusions of his day regarding the moon and the planets):

> "Moses wrote in a popular style things which, without instruction, all ordinary persons, endued with common sense, are able to understand; but astronomers investigate with great labor whatever the sagacity of the human mind can comprehend. Nevertheless, this study is not to be rejected, nor this science to be condemned, because some frantic persons are inclined boldly to reject whatever is unknown to them. For astronomy is not only pleasant, but also very useful to be known; it cannot be denied that this art unfolds the admirable wisdom of God."[2]

I think that we could include any and all of the modern categories of learning about God's world - physics, chemistry, geology, biology, ... - along with astronomy as "unfolding the admirable wisdom of God."

In the introduction to his commentary on the book of Genesis, speaking about the physical universe, Calvin wrote:

> "We know God, who is himself invisible, only through his works. .... men are commonly subject to these two extremes; namely, that some, forgetful of God, apply the whole force of their mind to the consideration of nature; and others, overlooking the *works* of God, aspire with a foolish and insane curiosity to inquire into his *Essence*. Both labor in vain. To be so occupied in the investigation of the secrets of nature, as never to turn the eyes to its Author, is a most perverted study; and to enjoy everything in nature without acknowledging the Author of the benefit, is the basest ingratitude. .... As for those who proudly soar above the world to seek God in his unveiled essence, it is impossible but that at length they should entangle themselves in a multitude of absurd figments. For God - by other means invisible - (as we have already said) clothes himself, so to speak, with the image of the world, in which he would present himself to our contemplation. .... Therefore, as soon as the name of God sounds in our ears, or the thought of him occurs to our minds, let us also clothe him with this most beautiful ornament; finally, let the world become our school if we desire rightly to know

God."[3]

So I try to avoid the two extremes mentioned by John Calvin, and, instead, acknowledge and praise the God of the Scriptures as the creator and governor of the universe, while at the same time accepting the well-supported results of the scientific study of God's world as his revelation of himself in his creation.

The early Christian theologian Augustine also wrote about the teachings of Scripture and our understanding of the physical world that results from our scientific study of that world. In discussing the first chapter of Genesis and God's work in creation, Augustine wrote:

"Usually, even a non-Christian knows something about the earth, the heavens, and the other elements of this world, about the motion and orbit of the stars and even their size and relative positions, about the predictable eclipses of the sun and moon, the cycles of the years and the seasons, about the kinds of animals, shrubs, stones, and so forth, and this knowledge he holds to as being certain from reason and experience. Now, it is a disgraceful and dangerous thing for an infidel to hear a Christian, presumably giving the meaning of Holy Scripture, talking nonsense on these topics; and we should take all means to prevent such an embarrassing situation, in which people show up vast ignorance in a Christian and laugh it to scorn. The shame is not so much that an ignorant individual is derided, but that people outside the household of the faith think our sacred writers held such opinions, and, to the great loss of those for whose salvation we toil, the writers of our Scripture are criticized and rejected as unlearned men. If they find a Christian mistaken in a field which they themselves know well and hear him maintaining his foolish opinions about our books, how are they going to believe those books in matters concerning the resurrection of the dead, the hope of eternal life, and the kingdom of heaven, when they think their pages are full of falsehoods on facts which they themselves have learnt from experience and the light of reason? Reckless and incompetent expounders of Holy Scripture bring untold trouble and sorrow on their wiser brethren when they are caught in one of their mischievous false opinions and are taken to task by those who are not bound by the authority of our sacred books. For then, to defend their utterly foolish and obviously untrue statements, they will try to call upon Holy Scripture for proof, and even recite from memory many passages which they think support their position, although *they understand neither what they say nor the things about which they make assertion.*"[4]

This paragraph is followed by a reference to I Timothy 1:7. In the first chapter of Paul's first letter to Timothy, the apostle Paul is urging Timothy to "command certain men not to teach false doctrines any longer," (verse 3) and then follows with verse 7: "They want to be teachers of the law, but they do not know what they are talking about or what they so confidently affirm."

In the above passage, Augustine expresses respect for the results of the scientific study of God's world, and he uses very blunt language to reprove Christians who (mis)use Scripture in attempting to oppose the valid and widely accepted scientific understanding of the physical world. He is especially emphatic in urging Christian believers to avoid driving unbelievers away from Christian faith by (mis)using Scripture to defend ideas which are contradictory to the well-established results of scientific study.

Recognizing that there may be some disagreements among Christians regarding interpretation of Scripture passages which deal with matters of creation and the physical world, Augustine devotes some space, especially in Chapter 18, to a discussion of how we should approach our understanding of Scripture in such passages. He wrote:

"In matters that are obscure and far beyond our vision, even in such as we may find treated in Holy Scripture, different interpretations are sometimes possible without prejudice to the faith we have received. In such a case, we should not rush in headlong and so firmly take our stand on one side that, if further progress in the search of truth justly undermines this position, we too fall with it. That would be to battle not for the teaching of Holy Scripture but for our own, wishing its teaching to conform to ours, whereas we ought to wish ours to conform to that of Sacred scripture."[5]

Stated in other words, Augustine advises us not to tie our Christian faith too tightly to the current (or an outdated) scientific understanding of the physical universe, because our knowledge and understanding of the physical world may undergo some changes as we learn more by scientific investigation. Building the structure of our Christian faith on a foundation of the scientific understanding of the world at any particular time in history is like building on the proverbial foundation of sand; our understanding of the world shifts with time. As was noted before, Romans 10:9 says: "if you confess with your mouth, 'Jesus is Lord,' and believe in your heart that God raised him from the dead, you will be saved."

Personal Likes and dislikes

There may be some results of our scientific investigation of God's universe that are not to your personal liking. Some of the explanations that have been supported by evidence from observations, and that have been widely accepted by scientists as valid, may conflict with what you thought the universe was like, or may be different from the traditional understanding of the Christian Scriptures that you had been comfortable with. Those of us who grew up in the context of the perspective that the Earth was created in six 24-hour days, not more than several thousand years ago, may be unhappy about the scientifically-based conclusion that Earth is billions of years old. Those of us who have been taught in Sunday School or elsewhere that living creatures were formed instantaneously by God's creation decree may be unhappy about the scientifically-based conclusion that there is a long history of change and development in the various living creatures that inhabit the Earth.

Some of us may want to accept some of the conclusions of modern scientific study, those that improve our creature comforts, or our economic welfare, or our health, and reject those that we personally dislike. But we don't really have that choice. Modern science comes to us as the total package.

Along with the discovery of deposits of mineral ores and energy fuels below the surface of the Earth, we learn that the rocks have been formed over long periods of time by the same kinds of processes that are in operation today. Along with our knowledge that the disease of poliomyelitis is caused by a virus that we can identify and isolate so that vaccines can be developed to prevent the disease, we learn that some viruses undergo significant and rapid genetic change in response to anti-viral drugs in their environment, thus lending support to the idea that living organisms change by an evolutionary process of natural selection. Along with the opportunities for improvement of health and improved approaches to the prevention of disease that are presented to us with our recently acquired detailed knowledge of the human genome, we learn that various sections of the human genome are strikingly identical to the genomes of other mammals, particularly chimpanzees, providing powerful evidence in support of the idea of common ancestry of humans and other mammals.

So the desired result and the disliked result come together, inseparable. The universe is what it is, whether we understand it or not. To the extent that we understand it, the universe is what we find it to be, whether we like it or not.

Some people might think that they would like to return to an earlier time, a time before all those scientific studies and results. Even if we could - and we cannot - I don't think any of us would like it. Smallpox and polio rampant; I remember the days before polio vaccines, and I don't think any of us want to return to that former time. The disease struck mostly young people, and mostly in late summer months. I remember the concern in my parents' eyes and voices as the news came that the son or daughter of a neighbor, or of a relative, had contracted polio, with its crippling and sometimes fatal consequences; we don't want that.

Some people think that we could praise God for his creation and government of the universe more fully, or more eloquently, if we didn't have all that scientific information about his world. But I think the opposite is true; we appreciate God all the more when we understand the beauty and intricate design in his universe in greater detail. Do you classical music lovers think that we would appreciate a great symphony more if we understood less about music? Of course not!

The task of the Christian church

The early Christian church, established within the Jewish tradition, had a difficult time accepting the idea that the gospel should be extended to Gentiles. Peter was criticized for even entering the home of the Roman Centurion, Cornelius. The apostles convened an international synod (convention) to examine the actions of Paul and Barnabas in baptizing Gentiles into the faith (Acts 15). Some of those early followers of Jesus Christ, being of Jewish origin, thought in traditional terms of God's people being part of the Jewish nation, and they taught that Gentile Christians should be required to observe Jewish customs and Old Testament ritual laws. The apostles, the leaders of the early Church, quoted various Old Testament Scriptures and interpreted (re-interpreted?) them to demonstrate that all people on Earth are the targets of God's love. In his Commentary on Acts of the Apostles, L.T. Johnson emphasized that this action was taken by the Church, and that it was not merely the action of an individual apostle. Johnson points out that God

> "subtly but surely uses the apostles' statements to shape a new definition of 'the people of God' as one based on messianic faith rather than on ethnic origin or ritual observance. He establishes as a fundamental principle that the Church's responsibility is not to dictate God's action, but to discern it; not to close the Scriptures to further interpretation, but to open them."[6]

The people of God face a somewhat similar situation in the world today with regard to the results of scientific study of God's world. Those results are being opposed by some Christians on the basis of traditional understanding of God's Word in the Scriptures. Scripture passages are being quoted to "prove" that the results of that scientific study cannot possibly be correct. But scientific study is the investigation of God's handiwork in the physical world. In this area of study and discovery, too, we - individual Christians and Christian churches - would do well, I think, to remember and apply what Johnson calls the "fundamental principle" that the Christian's, and the Christian church's, responsibility "is not to dictate God's action, but to discern it; not to close the Scriptures to further interpretation, but to open them."

References

[1] John Calvin, *Institutes of the Christian Religion* (1559) (Translation by Henry Beveridge, 1845), (Grand Rapids: Wm. B. Eerdmans Publishing Co., 1957), Book I, Chapter V, paragraph 2, 51-52.

[2] John Calvin, *Commentary on the First Book of Moses* (1551) (Translation by John King, 1847) (Grand Rapids, MI: Baker Book House, 1948), 86.

[3] Calvin, *Commentary*, 59-60.

[4] Augustine, *The Literal Meaning of Genesis* (415 A.D.) (Translation by John Hammond Taylor) (New York: Newman Press, 1982), Book One, Chapter 19, Paragraph 39, 42-43.

[5] Augustine, *Literal*, Book One, Chapter 18, Paragraph 37, 41.

[6] L.T. Johnson, *The Acts of the Apostles* (Sacra Pagina Series, Daniel J. Harrington, Ed., Vol. 5), (Collegeville, Minnesota: The Liturgical Press, 1992), 280.

In the end, the truth will win.

Suggested Reading

As you read any or all of these suggested reading materials, you might react to some idea or thought with the response, "I disagree!" You might react to some idea or thought with the response, "I'm not sure about that; let's study and consider that some more before I make up my mind." And I expect that you will react to some ideas or thoughts with a hearty "Amen!" And thus we grow in our understanding, and also in our commitment to Jesus Christ as Lord of our lives and Master of the universe.

Always, as we read and study, we work from the foundation of our faith in Jesus Christ and his resurrection, and we maintain our trust in his promises. As the quip above my desk reads, "Keep an open mind, but not so open that your brains fall out."

Here are the suggestions for further reading:

Francis S. Collins, *The Language of God*, New York: Free Press (Simon & Schuster, Inc.), 2006.

Deborah B. and Loren D. Haarsma, *Origins*, Grand Rapids: Faith Alive Christian Resources, 2007.

Denis Lamoureux, *Evolutionary Creation*. Eugene, OR: Wipf and Stock, 2008.

Howard J. Van Till, *The Fourth Day*. Grand Rapids: Eerdmans Publishing Company, 1986.

Howard J. Van Till, Robert E. Snow, John H. Stek, and Davis A. Young. *Portraits of Creation*, Grand Rapids: Eerdmans Publishing Company, 1990.

Howard J. Van Till, Davis A. Young, and Clarence Menninga. *Science Held Hostage*, Downers Grove, IL: Intervarsity Press, 1988.

John H. Walton, *The Lost World of Genesis One*, Downers Grove, IL: Intervarsity Press, 2009.

Davis A. Young, *The Biblical Flood*, Grand Rapids: Eerdmans

Publishing Company, 1995.

Davis A. Young and Ralph F. Stearley. *The Bible, Rocks, and Time*, Downers Grove, IL: Intervarsity Press, 2008.

Appendix A
Second Law of Thermodynamics (Chapter 13)

The basic ideas of thermodynamics have been presented in ordinary prose in Chapter 13, and that discussion need not be repeated here. We will, however, provide some numerical information and some mathematical formulas to demonstrate that some spontaneous processes occurring on Earth result in a decrease in the entropy of local, open systems.

The Second Law of Thermodynamics

The Second Law of Thermodynamics, applied to chemical reactions, is presented in the college chemistry textbook by Brown, LeMay, Bursten and Burdge, in a chapter entitled "Chemical Thermodynamics."

The important concepts that are presented in that chapter are summarized as follows:

    1. "In this chapter we will put a more formal stamp on our understanding that changes that occur in nature have a directional character: They move *spontaneously* in one direction but not in the reverse direction.

    2. "The thermodynamic function, *entropy*, is a state function (like enthalpy and internal energy), which may be thought of as a measure of disorder or randomness.

    [Note by CM: A "state" function depends only on the initial and final states of the substance being described, regardless of the path that is taken in proceeding from initial state to final state.]

    3. "The *second law of thermodynamics* tells us that in any spontaneous process the net entropy of the universe (system plus surroundings) increases.

    4. "The *third law of thermodynamics* states that the entropy of a perfect

crystalline solid at 0 K(absolute) is zero. From this reference point the absolute entropies of pure substances at temperatures above absolute zero can be calculated from experimental data.

5. "The *free energy* (or *Gibbs free energy*) is a measure of how far removed a system is from equilibrium. It measures the maximum amount of useful work obtainable from a given process and yields information on the direction in which a chemical system will proceed spontaneously."[1]

We will apply the principles of thermodynamics to the examples of local systems that were listed in Chapter 13 as systems that experience a decrease in entropy as a result of spontaneous, natural processes. To do that, we must first provide some definitions and explanations of symbols that will be used in the discussion:

NOTE: The change in some thermodynamic property, such as entropy, is generally represented in shorthand fashion by the fourth letter of the Greek alphabet, an upper-case "delta," thus "Δ."

1. The chemical elements are represented by the common chemical symbols. The following few elements will be considered in this discussion:

C = carbon; H = hydrogen; Fe = iron; Mg = magnesium; O = oxygen

2. Other symbols and their meanings:

a. Entropy is represented by S; standard entropy by $S°$; change in entropy by ΔS.

b. Energy is measured in joules, represented by J, or in calories. 1 calorie = 4.184 joules

c. Enthalpy (heat) is represented by H; standard enthalpy by $H°$; change in enthalpy by ΔH, and is measured in calories.

d. Free energy is represented by F; standard free energy by $F°$; change in free energy by ΔF. (Some publications refer to "free energy" as "Gibbs free energy" in honor of J. Willard Gibbs, American chemist/mathematician who developed the relationship, and use the symbol "G" instead of "F.")

e. The unit of measurement of the amount of a substance involved in a reaction is the "mol" (pronounced "mole"). One mol of a substance is the number of grams that equals the molecular weight of that substance (or atomic weight for mono-atomic elements). The atomic weights of the elements listed in (1) are: C = 12; H = 1; Fe = 56; Mg = 24; O = 16.

(1). Thus, the atomic weight of Fe is 56; one mol of Fe = 56 grams of iron.

(2). The molecular weight of carbon dioxide, CO_2, is the sum of all the atomic weights in the formula, as follows:

12 + (2 x 16) = 44; one mol of CO_2 = 44 grams of carbon dioxide.

f. Temperature is measured in degrees Kelvin, (or, in older literature, degrees Absolute), represented in thermodynamic data and calculations by K (upper case K).

(1). A difference of one degree on the Kelvin scale is the same amount of temperature difference as one degree on the Celsius scale (or, in older literature, the Centigrade scale).

(2). Absolute zero = - 273.16°C; thus, K = °C + 273.16

g. Tabulated values of various thermodynamic quantities are generally expressed as "standard" quantities, as in "standard entropy." "Standard" means under the conditions of one atmosphere of pressure and a temperature of 298.16 K = 25° C.

Thermodynamic properties of various materials

The thermodynamic properties of a large number of materials have been measured using well-insulated apparatus called a calorimeter. These thermodynamic properties have been published in the scientific literature, and have been collected and published in "tables" of thermodynamic properties. Such tables may be found in publications like *Handbook of Chemistry and Physics*, 47th Edition.[2] The tables in that publication are excerpted from the larger publication "Selected values of Chemical Thermodynamic Properties," *Circular of the National Bureau of Standards 590*, issued in 1952.

Examples described in Chapter 13

Now we will proceed to put some numbers with the examples from Chapter 13 of processes that proceed spontaneously in nature, processes that result in a <u>decrease</u> in the entropy of the local, open system.

1. Methane(g) burns spontaneously in oxygen(g) to form carbon dioxide(g) and water(*l*).

NOTE: the expression in parentheses is the state of matter of that substance at standard conditions: 1 atmosphere of pressure and 25°C; (s) is solid, (*l*) is liquid, (g) is gas.

First, we write a balanced chemical equation for the process:

$$CH_4(g) + 2\,O_2(g) \rightarrow CO_2(g) + 2\,H_2O(l)$$

Next, we consult the tables in the *Handbook* to get the standard entropy, $S°$, for each substance involved in the reaction

$S°_{(methane)} = 44.50$ entropy units (e.u.)

$S°_{(oxygen)} = 49.003$ e.u.

$S°_{(carbon\ dioxide)} = 51.06$ e.u.

$S°_{(water)} = 16.716$ e.u.

Next, we determine the change in entropy, ΔS, given by

$$\Delta S = S°_{(products)} - S°_{(reactants)}$$

The standard entropy values are given *per mol*, so we must take the coefficients of the balanced chemical equation into account.

$\Delta S = S°_{(carbon\ dioxide)} + 2\,S°_{(water)} - S°_{(methane)} - 2\,S°_{(oxygen)}$

$= 51.06$ e.u. $+ 2(16.716)$ e.u. $- 44.50$ e.u. $- 2(49.003)$ e.u.

$= -58.014$ e.u.

So, as you see, the result is a negative number, meaning that the spontaneous burning of methane in oxygen results in a <u>decrease</u> in the entropy of the local, open system.

The Second Law of Thermodynamics states that the entropy of the universe increases as a result of a spontaneous process, so the entropy of the surroundings must increase to a greater extent than the decrease in the entropy of the local, open system when methane burns in oxygen. We will take a look at that aspect of the process a bit later.

2. Many high school and college chemistry experiments include igniting a ribbon of magnesium, and watching it burn with intense white light. Magnesium is a component of many fireworks. Magnesium(s) burns spontaneously in oxygen(g) to form magnesium oxide(s).

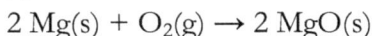

$2 Mg(s) + O_2(g) \rightarrow 2 MgO(s)$

$S°_{(magnesium)} = 7.77$ e.u.

$S°_{(magnesium\ oxide)} = 5.8$ e.u.

$S°_{(oxygen)}$ was given above $= 49.003$ e.u.

$\Delta S = S°_{(magnesium\ oxide)} - S°_{(magnesium)} - S°_{(oxygen)}$

$\Delta S = 2(5.8)$ e.u. $- 2(7.77)$ e.u. $- 49.003$ e.u.

$= - 52.943$ e.u.

As you see, the entropy of the local, open system decreases when magnesium burns spontaneously in oxygen.

3. Hydrogen(g) burns in oxygen(g) to form water(l)

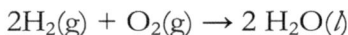

$2H_2(g) + O_2(g) \rightarrow 2 H_2O(l)$

$S°_{(hydrogen)} = 31.211$ e.u.

$\Delta S = 2 S°_{(water)} - 2 S°_{(hydrogen)} - S°_{(oxygen)}$

$= 2(16.716)$ e.u. $- 2(31.211)$ e.u. $- 49.003$ e.u.

$= - 77.993$ e.u.

Again, the entropy of the local, open system decreases when hydrogen burns spontaneously in oxygen.

4. Iron(s) rusts spontaneously in the presence of oxygen(g), as we all know very well.

$4 Fe(s) + 3 O_2(g) \rightarrow 2 Fe_2O_3$

$S°_{(iron)} = 6.491$ e.u.

$S°_{(ferric\ oxide)} = 21.5$ e.u.

$\Delta S = 2 S°_{(ferric\ oxide)} - 4 S°_{(iron)} - 3 S°_{(oxygen)}$

$= 2(21.5)$ e.u. $- 4(6.491)$ e.u. $- 3(49.003)$ e.u.

$= - 129.973$ e.u.

Yes, even the rusting of iron, so frequently used as symbolic of the decay and deterioration of our world, results in a decrease in the entropy of the local, open system.

Many other examples could be given, but these suffice to demonstrate that local, open systems can experience a decrease in entropy in spontaneous, irreversible processes. If truth be known, there is no valid basis for the claim that a decrease in the entropy of local, open systems by natural, spontaneous processes would be contradictory to the Second Law of Thermodynamics.

The Universe

The Second Law of Thermodynamics states that a spontaneous process always results in an increase in the entropy of the universe. The universe, of course, consists of the local, open system _and_ its surroundings. In the cases considered above, the entropy of the local, open system _decreases_, so the entropy of the surroundings must have increased by an even greater amount.

So, is there some way to account for the entropy of _both_ a local, open system _and_ its surroundings? Yes, there is. It is called "free energy," or "Gibbs free energy." The mathematical expression that defines free energy is as follows:

$F = H - TS$ [equation 1]

where H is the enthalpy, T is the absolute temperature, and S is the entropy.

For a process occurring at constant temperature, the change in free energy of the system, ΔF, is given by the expression

$\Delta F = \Delta H - T\,\Delta S$ [equation 2]

We should note that the surroundings of any local, open process serve essentially as a very large, constant temperature heat source, or heat absorber. For a reaction or process that occurs at constant pressure,

$\Delta S_{(surroundings)} = (-\,\Delta H_{(system)})/T$ [equation 3]

NOTE: Equation 3 is derived from the potential amount of work that may be accomplished by the process being considered. See reference 1 for details.

By definition

$\Delta S_{(universe)} = \Delta S_{(system)} + \Delta S_{(surroundings)}$ [equation 4]

Substituting the right side of equation 3 for $\Delta S_{(surroundings)}$, we have

$\Delta S_{(universe)} = \Delta S_{(system)} + (- \Delta H_{(system)})/T$ [equation 5]

Note that the change in entropy of the universe is now defined by two terms for the local system, and the temperature. The terms for the local system are measureable quantities, and the temperature at standard conditions is always 298.16 K.

We can rearrange terms by multiplying both sides of equation 5 by - T, so that

$- T \Delta S_{(universe)} = \Delta H_{(system)} - T \Delta S_{(system)}$ [equation 6]

Note that the right side of equation 6 is identical to the right side of equation 2 when applied to the local system. Thus, for a process occurring at constant temperature and constant pressure, the free energy change in the local system is given by

$\Delta F_{(system)} = - T \Delta S_{(universe)}$ [equation 7a]

By multiplying both sides of equation 7a by -1, we have

$- \Delta F_{(system)} = T \Delta S_{(universe)}$ [equation 7b]

Usefulness of free energy relationships

The change in free energy for the local, open system tells us something very useful about whether or not a chemical reaction, or any natural process, is spontaneous. The Second Law of Thermodynamics states that any spontaneous (irreversible) reaction produces an <u>increase</u> in the entropy of the universe. Consequently, any spontaneous, irreversible reaction also produces a <u>decrease</u> in the free energy of the local, open system. Or, in other words, if ΔF for the process is negative, the process is spontaneous; if ΔF is positive, the process is not spontaneous. Conversely, if a process is spontaneous, ΔF will be negative.

So, if we take another look at equation 2, applied to the local system, we have

$\Delta F_{(system)} = \Delta H_{(system)} - T \Delta S_{(system)}$

We note that the right hand side of the equation consists of two terms. At any temperature above absolute zero, T is positive. The enthalpy change in a specific process or reaction may be positive or negative, and the entropy change may be positive or negative. Thus we have the following range of possibilities for the sign of ΔF:

1. If ΔH is negative and ΔS is positive, ΔF will be negative, and the process is spontaneous.

2. If ΔH is negative and ΔS is negative, the sign of ΔF depends on the relative magnitudes of ΔH and $T \Delta S$.

3. If ΔH is positive and ΔS is negative, ΔF will be positive, and the process is not spontaneous.

4. If ΔH is positive and ΔS is positive, the sign of ΔF depends on the relative magnitudes of ΔH and $T \Delta S$, as in case 2.

We should also note that the quantities on the right hand side of equation 2 refer to the local, open system and therefore can be determined by experiment. Thus, the free energy change in the process depends on quantities that apply to the local system, and also yields information on the entropy change in the universe.

The change in enthalpy for the process or reaction by which a substance is formed from its constituent elements can be measured; this quantity is called the "enthalpy of formation" of that substance, represented by the symbol ΔH_f. The change in free energy for the reaction or process by which a substance is formed from its constituent elements can be measured; this quantity is designated the "free energy of formation" of that substance, represented by the symbol ΔF_f. Under the conditions of 1 atmosphere of pressure and a temperature of 25° C, the standard enthalpy of formation and the standard free energy of formation for various substances have been determined by experiment, and have been published in the *Handbook* and other references. Because we are interested in differences rather than absolute values, the standard enthalpies of formation and the standard free energies of formation of the pure elements in their standard states are set to zero.

We can then determine the free energy change for any reaction involving those substances whose standard free energy of formation is known. Let us do that for one of the examples whose change in entropy was considered above, that is, for the process of rusting of iron, using the values for standard free energy found in the *Handbook*:

$$4 \text{ Fe(s)} + 3 \text{ O}_2(g) \rightarrow 2 \text{ Fe}_2\text{O}_3(s)$$

$$\Delta F_{(system)} = \Delta F_f^{\circ}{}_{(products)} - \Delta F_f^{\circ}{}_{(reactants)}$$

$$= 2 \Delta F_f^{\circ}{}_{(ferric\ oxide)} - 4 \Delta F_f^{\circ}{}_{(iron)} - 3 \Delta F_f^{\circ}{}_{(oxygen)}$$

$$= 2 \text{ mol } (-177 \text{ calories/mol)} - 0 - 0$$

$$= -354 \text{ calories}$$

The free energy change is negative, which tells us that the reaction, as written, is spontaneous. While there is a decrease in entropy for this local, open system, as we noted earlier, the entropy of the surroundings increases by a greater amount, so that there is a net increase in the entropy of the universe.

As we noted in equation 7b above, the negative of the free energy of a local, open system is equal to the product of the absolute temperature, T, and the change in the entropy of the universe, ΔS. Since T is always positive for any temperature above absolute zero, the change in the entropy of the universe will be positive for any process in which the change in free energy, ΔF, is negative.

As we noted in prose in Chapter 13, there is no valid basis for the claim that a decrease in the entropy of all local, open systems by natural, spontaneous processes would be contradictory to the Second Law of Thermodynamics.

References

[1]T.L. Brown, H.E. LeMay, Jr., B.E. Bursten, and J.R. Burdge, *Chemistry, The Central Science*, Ninth Edition, (Upper Saddle River, NJ: Prentice Hall, 2003), 735.

[2]Robert C. Wiest, Editor-in-Chief, *Handbook of Chemistry and Physics*, 47th Edition, (Cleveland: The Chemical Rubber Company, 1962), Section D, 32-51.

[3]Wiest, *Handbook*, D-38.

[4]Wiest, *Handbook*, D-22.

Appendix B
Radioactivity and Radioactive Decay (Chapter 16)

The basic characteristics of radioactive isotopes and radioactive decay have been presented in Chapter 16. Additional information, some of it a bit more technical, is presented in this Appendix for those readers who want a bit more detail.

Rate of Decay

A graph depicting the phenomenon of radioactive decay is found in Chapter 16, along with a narrative that describes in words what the graph depicts. We should note, in addition, that the description of decay is a statistical description. Even if we could label each atom of a radioactive isotope in a sample, we would not be able to predict precisely when a particular atom would be transformed in the decay process. If there is a large number of atoms of a radioactive isotope in our sample, however, we know that, on average, half of them will decay in one half-life of time. That prediction is based on the average of a large number of measurements, and each individual measurement may give a result that is somewhat above or below the average. It is somewhat similar to tossing a coin, or throwing dice. If we toss a six-sided die a number of times, we would expect the number "six" to come up once in every six tries; in actual practice, with a limited number of tries, "six" might come up somewhat more often, or less often, than the average of many tries. Because of this probabilistic nature of radioactive decay, there is always an uncertainty associated with each measurement. In typical practice, the uncertainty of measurements reported in the scientific literature is not more than a few percent of the total, often less than one percent, and the amount of uncertainty should always be reported along with the published result.

Modes of decay

There is another aspect of radioactive decay that we should note in connection with this discussion, referred to as "modes" of radioactive decay. There are four different processes that together are called "radioactive decay." Most radioactive isotopes undergo decay by one of these processes; in some cases, the radioactive isotope undergoes "branching" decay, that is, some atoms of the isotope decay by one of these processes, some by another. The processes involved are taking place in the nucleus of the atom.

NOTE: These processes were initially grouped as "radiation" of three types, given Greek letters "alpha," "beta," and "gamma." It was later found that alpha and beta radiation involved emission of particles from the nucleus, but the Greek letter part of the earlier terminology continues to be in use.

The modes of decay are:

1. Alpha particle emission

a. An alpha particle consists of two protons and two neutrons. Having two protons identifies it as being identical to the nucleus of a helium atom, and the two protons plus two neutrons give it an atomic mass number of 4.

b. The daughter of a radioactive isotope that undergoes decay by emitting an alpha particle will have two fewer protons and two fewer neutrons in its nucleus than the parent isotope had. Therefore, it will have an atomic number of two units less than that of the parent, and an atomic mass number of four less.

c. For example: uranium is atomic number 92; the isotope U-238 undergoes radioactive decay by alpha particle emission, so the daughter isotope has atomic number 90, which is the element thorium, and the daughter has atomic mass number 234, which is the isotope Th-234.

2. Beta particle emission

We note that there are two types of beta particles, and thus two modes of decay by beta particle emission.

a. Beta-minus

i. A beta-minus particle has a negative electrical charge equal to that on an electron, and a mass identical to that of an electron; it is indistinguishable from an electron.

ii. The daughter of a radioactive isotope that undergoes decay by emitting a beta-minus particle will have lost a negative electrical charge from its nucleus equal to that on an electron, or, in other words, will have gained additional positive electrical charge equal to that on a proton; therefore it will have one more proton in its nucleus than that of the parent isotope. The mass of a beta particle is less than the mass of a proton by a factor of 1836, so the atomic mass number of the daughter is the same as that of the parent.

iii. For example, cobalt is atomic number 27; the isotope Co-60 undergoes decay by emitting a beta-minus particle, so the daughter has atomic number 28, which is the element nickel, and there is no change in atomic mass number, so the daughter is the isotope Ni-60.

b. Beta-plus

i. A beta-plus particle has a positive electrical charge equal to that on a proton, and a mass equal to that of an electron. It is sometimes called a "positron."

ii. The daughter of a radioactive isotope that undergoes decay by emitting a beta-plus particle will have lost a positive electrical charge from its nucleus equal to that on a proton; therefore it will have one less proton in the nucleus than the parent. Because of the small mass of the beta-plus particle, there is no change in atomic mass number.

iii. For example, aluminum is atomic number 13; the isotope Al-25 undergoes decay by emitting a beta-plus particle, so the daughter is atomic number 12, which is magnesium, and there is no change in atomic mass number, so the daughter is the isotope Mg-25.

3. Electron capture

a. In some cases, the nucleus of a radioactive isotope "captures" an electron from outside the nucleus, that is, the electron from outside the nucleus becomes a part of the nucleus.

b. The daughter of a radioactive isotope that undergoes decay by electron capture will have gained a negative electrical charge on its nucleus equal to that on an electron, or, in other words, will have lost a positive electrical charge equal to that on a proton. There will be no change in the atomic mass number. Note that this result is the same as that in the case of decay by beta-plus emission, although by a different process.

c. For example, iron is atomic number 27; the isotope Fe-55 undergoes decay by electron capture, so the daughter is atomic number 26, which is the element manganese, and there is no change in atomic mass number, so the daughter is the isotope Mn-55.

In most cases, these processes of radioactive decay are associated with the emission of gamma radiation from the nucleus. Gamma radiation, like x-rays, is a form of electromagnetic radiation with short wavelength.

Appendix C
Measuring Ages of Rocks (Chapter 17)

In this appendix I will provide a bit more information about argon being excluded from crystalline igneous rocks, describe hourglass analogies for the Potassium-40/Argon-40 and the Uranium-238/Lead-206 methods, and present an additional method of measuring ages of rocks not described in Chapter 17.

Argon in crystalline igneous rocks

The validity of the K-40/Ar-40 method depends on the condition that the sample contain no argon-40 when the rock was initially formed. So what is the basis for our conviction that this condition can be met in properly chosen samples?

When molten magma cools and solidifies slowly enough to form distinct mineral crystals, the atoms in these mineral crystals are arranged in a regular 3-dimensional geometric pattern. Argon is a "noble" element, that is, it does not ordinarily form compounds with other elements, and it is a gas at ordinary temperatures. Gases can be dissolved in liquids; for example, water at room temperature ordinarily contains some dissolved air. Even more of a gas can be dissolved in liquids if the mixture can be kept under pressure; for example, carbon dioxide is kept in solution in carbonated beverages under pressure, but it comes out of solution as bubbles of gas when the container is opened and the pressure is released. There probably are argon atoms dissolved in molten magma while it is under pressure deep underground, but there is no room for the argon atoms in the structures of the mineral crystals that form as the magma cools and solidifies. Therefore, the argon atoms are excluded from the crystals that make up crystalline igneous rock; there would be no argon in the rock when it was formed.

You can demonstrate the principle involved by doing an experiment in your kitchen. Dissolve about one-half teaspoon of salt in about eight ounces of water, and stir the mixture thoroughly. Then set the container in the freezer until about half of the liquid has frozen. (Use plastic rather than glass, because water expands when it freezes and might break the glass if it is accidentally allowed to freeze completely.) After about half of the liquid has turned to ice, pour the remaining liquid into a different container, then rinse the ice surface with a little bit of fresh water, and add the rinse water to the liquid portion. Now melt the ice, pour the meltwater into a clean pan, evaporate the water, and observe how much of the salt had been incorporated into the ice. In a separate pan, evaporate the water from the liquid fraction, and observe how much of the salt had remained in the liquid fraction.

CAUTION: Evaporate the last little bit of liquid slowly and carefully; if you heat it too strongly the concentrated salt solution will decrepitate (pop and spatter).

NOTE: A small amount of dissolved solids is found in all water supplies, so you might want to evaporate about four ounces of your tap water to find out how much solid residue is due to the tap water alone.

Water forms crystalline ice when it is cooled and solidified. The reasonable conclusion from your experiment is that the ice crystals do not have space available for the sodium and chloride ions that make up the salt, so nearly all of the salt is excluded from the ice crystals as the water freezes, and is left in the liquid water.

Similarly, there is no space for argon atoms in the mineral crystals formed in igneous rocks.

An hourglass analogy for K-40/Ar-40

In our hourglass analogy for the K-40/Ar-40 method, we need to build an hourglass that contains a specially prepared sand made up of a well-stirred mixture of 89% white sand grains and 11% red sand grains. The white sand grains represent those atoms of K-40 that will decay to Ca-40, and the red sand grains represent those that will decay to Ar-40. (All the sand grains are the same size and density, so they flow through the constriction of the hourglass in exactly the same ratio as they exist in the prepared mixture.) This sand mixture would be placed in the top of the hourglass when it begins running.

When we encounter our hourglass with some of the sand still in the top:

1. We measure the total amount of sand in the top of the hourglass at present: this corresponds to the amount of K-40 in the rock sample at present.

2. We cannot measure the amount of sand that was in the top of the hourglass when it began running; we weren't there. But we know that there were no red sand grains in the bottom when the hourglass began running, corresponding to our assurance that there was no Ar-40 in the rock sample when it was initially formed. (There may have been any number of white sand grains in the bottom when it began running, corresponding to the likelihood that there were some Ca-40 atoms in the rock sample when it was formed.) So we count the number of red sand grains in the bottom of the hourglass at the present time, and we know that for every 11 red sand grains in the bottom, 100 sand grains passed through the constriction in the hourglass since it began running. So we add the number of sand grains that have passed through the constriction to the number that are still in the top of the hourglass at present to get the number of sand grains that were in the top when the hourglass began running.

3. We determine the rate at which the sand flows through the constriction in the hourglass, and then we can calculate how long the hourglass has been running. The rate at which the sand flows through the narrow constriction represents the rate of decay of K-40, with its half-life of 1.3 billion years.

Now we can calculate how long the hourglass had been running when we found it: this corresponds to the calculation for determining the age of a rock sample with the K-40/Ar-40 procedure.

An hourglass analogy for the U-238/Pb-206 method

In our hourglass analogy for the U-238/Pb-206 system we will construct two hourglasses, labeled "#1" and "#2". In the bottoms of both hourglasses we will place a mixture of white and red sand grains; for this example, we will choose a ratio of 12 white sand grains to 1 red sand grain. The ratio of white to red sand grains represents the ratio of the amounts of common Pb-206 to Pb-204. The ratio of common Pb-206 to Pb-204 found in actual rock samples changes with time in actual Earth history, so the choice of 12 to 1 in this example is only for the purpose of illustration. In the top of hourglass #1 we have only white sand grains, representing U-238; in the top of hourglass #2 there is no sand at all.

After the hourglasses have been in operation for some time, we determine how long they have been in operation as follows:

1. We measure the amount of white sand grains in the top of hourglass #1 at the present time. This represents the amount of U-238 in the rock sample at the present time.

2. We cannot measure the amount of white sand grains that were in the top of hourglass #1 when the hourglass began running; we weren't there at the time. We know, however, that the ratio of white to red sand grains in the bottom of hourglass #1 increases as white sand runs through the constriction in the hourglass. We also know that the ratio of white to red sand grains in the bottom of hourglass #1 when it began running is the same as that in the bottom of hourglass #2. We also know that the ratio of white to red sand grains in the bottom of hourglass #2 has not changed with time - no white sand grains have been added. So we measure the ratio of white sand grains to red sand grains in hourglass #2, and we measure the ratio of white sand grains to red sand grains in hourglass #1 at the present time. We subtract the ratio in #2 from the ratio in #1, and the difference tells us how many white sand grains have been added since

the hourglass began running. We add that number to the number of white sand grains in the top of hourglass #1 at the present time: the sum is the number of white sand grains that were in the top of hourglass #1 when it began running. This corresponds to the amount of U-238 in the rock sample when the rock was formed.

3. We measure the rate at which sand grains run through the constriction in hourglass #1; this corresponds to the rate at which U-238 decays to Pb-206.

Now we can calculate how long the hourglass has been running. It's a little more involved than the simpler hourglass that we constructed as an analogy for K-40/Ar-40, but it would work for measuring the passage of time.

Spontaneous fission of uranium-238

There is another phenomenon in rocks containing uranium that can be used to measure the age of the rock. This method does not involve ordinary radioactive decay, and is not concerned with daughter isotopes. It is the spontaneous fission of U-238.

Nuclear fission of uranium-235 in a rapidly growing chain reaction is the source of the energy in nuclear explosives. Nuclear fission in a sustained and controlled chain reaction is the source of energy in nuclear power reactors, now familiar to us in our "nuclear age" of history. Such chain reaction fission is induced by interaction with thermal (low energy) neutrons.

Another type of nuclear fission, called spontaneous fission, takes place without any external interactions in nuclei with atomic mass number higher than about 230. The rate at which such spontaneous fission occurs can be measured. Most of the radioactive decays of uranium-238 atoms occur by alpha particle emission, but there are 5 spontaneous fissions for every 10 million decays by alpha particle emission. Another way to describe that rate would be to say that in each gram of U-238 there are 25 spontaneous fissions per hour.

The nuclear fission of a heavy element such as uranium produces two fragments of unequal mass, typically one with about two-thirds the mass of the heavy element nucleus, and the other with about one-third the mass of the heavy element nucleus. Since both of these nuclei are positively charged, they are strongly repelled by each other, and they speedily move away from each other in opposite directions, each carrying a large positive electrical charge and a great deal of kinetic energy.

As the fission fragments travel through matter, they are slowed by interactions with the atoms they encounter along the way; as they slow down, they pick up electrons from their surroundings, and eventually they come to a stop as neutral atoms. If the fission occurs within solid material, these fission fragments leave a trail

Figure 8. Spontaneous fission tracks in zircon.
Courtesy of J.D. Macdougall

of damage in their wake, caused by their high electrical charge and their kinetic energy as they disturb the electron structure of the atoms they collide with. In a crystalline solid the damage disturbs the geometric structure of the crystal as well. Such damage trails are very tiny, and they are not readily visible, even with a good microscope.

However, etching a freshly polished surface of a mineral crystal with an appropriate acid will preferentially attack the material in the damage trails and enlarge them. Now called "fission tracks," they are readily visible under an optical microscope, as, for example, in a sample of the mineral zircon shown in Figure 8.[1] (The fission tracks in Figure 8 are magnified 3100 times.)

As years go by, more and more atoms of U-238 will fission spontaneously and leave damage trails in any mineral crystal that contains uranium. To determine the age of the rock by this method, we need to measure the concentration of uranium-238 in the mineral, and count the fission tracks in a measured area of the polished and etched surface of the mineral sample. Knowing the rate at which spontaneous fission occurs in uranium-238, we can then calculate the age of the rock.

The mineral zircon often occurs as a minor, or accessory, mineral in igneous rocks. Uranium atoms are about the same size as zirconium atoms, and uranium atoms are commonly found in zircon crystals as replacements for some of the atoms of zirconium in the mineral. The fission track method is useful for measuring the ages of such rocks.

Some further comments

There are some details to attend to in doing the sorts of measurements I have outlined; proper equipment is needed, proper procedures for preparing the sample must be used, etc. As noted before, it is important to have some assurance that none of the parent isotope has been added to or lost from the sample during its history, and that none of the daughter isotope has been added to or lost from the sample during its history, except for the processes of radioactive decay.

Some understanding of the behavior of solid materials is important for proper sample selection. The elements in rocks may be able to move within the solid material at elevated temperatures, even if the minerals in the rock are not hot enough to be melted. This is especially true for argon, which is a gas at ordinary temperatures, and does not enter into combination with other elements to form chemical compounds. Such movement of parent or daughter isotopes would make a radiometric age measurement invalid. For example, we know that metamorphic rocks are formed when previously existing rocks are subjected to high temperatures, and often to high pressures as well. A measurement of K-40/Ar-40 in such samples would not tell us the age of the rocks that existed prior to their metamorphism, because it is very likely that the Ar-40 moved, probably escaping from the rock entirely, during the period of high temperature that produced the metamorphism. However, the method might be useful for telling us how long ago the rock cooled to normal temperatures following the metamorphism; in referring to such an event, we say that the radiometric clock was "reset" by metamorphism.

We noted in Chapter 17 that sedimentary rocks do not provide reliable samples for radiometric age measurement. The ages of fossil remains of living organisms is of considerable interest to us, but since fossils of living organisms are most often found in sedimentary rocks, we are not normally able to measure the age of the fossils directly. However, we can often "bracket" the age of a sedimentary deposit by the radiometric ages of crystalline igneous rocks or of volcanic rocks that were deposited within or above or below the sedimentary layer, and thus learn the ages of the fossils.

For a more detailed discussion of these methods, and for information on a wider range of radiometric age determination methods, I suggest that you consult a book on radiometric age measurement.[2,3]

References

[1]J.D. Macdougall, "Fission Track Dating," *Scientific American*, 235, No. 6 (1976), 114-22.

[2]Henry Faul, *Ages of Rocks, Planets and Stars*, (New York: McGraw-Hill

Book Co., Inc., 1966).

[3]Brent Dalrymple, *The Age of the Earth*, (Stanford: Stanford University Press, 1991).

Appendix D
Constancy of Decay Rates (Chapter 19)

A number of experiments have been performed to examine the question of the constancy of radioactive decay rates. In some cases certain changes in chemical structure or external pressure were found to have a small effect on radioactive decay rate. To understand what is going on in those cases we must consider each of the various "modes" of radioactive decay, presented in Appendix B.

All of these processes take place in the nucleus of the radioactive atom. Most radioactive isotopes decay by only one of these processes, though some, like potassium-40, undergo "branching" decay, some atoms of the isotope decaying by one process and some by another. Each of these modes of decay will result in a change in the electrical charge on the nucleus, producing a daughter isotope which is a different element from the parent.

Now, the important questions are as follows: 1) Which, if any, of these processes can be affected by factors which might result in a change in the rate of decay? 2) Do any of the isotopes used in measuring the ages of rocks undergo radioactive decay by a process(es) in which a change in the rate of decay might occur because of such external factors? 3) If a radioactive isotope used in measuring the ages of rocks is subject to a change in decay rate with changes in surrounding factors, how large a change in decay rate has been observed, or might reasonably be expected?

No change in the decay rate of a radioactive isotope decaying by alpha particle emission or beta particle emission has ever been observed. A review article by G.T. Emery reports early attempts to test possible changes in decay rates in the "Introduction" of that article.[1] With our present understanding of the structure of atoms, including their nuclei, no change in the decay rate of an isotope decaying by alpha or beta particle emission is expected under any feasible changes in the environment in which rocks exist on Earth. (In the interior of hot stars, where most or all of the electrons have been stripped away from atomic nuclei, other considerations may be important, but rocks on Earth don't exist in such environments.) Nearly all of the radioactive isotopes used to measure ages of rocks undergo decay by alpha and/or beta particle emission. Measuring ages of rocks by methods using those isotopes would not be expected to be in error because of changes in decay rates due to changes in external environment.

The process of electron capture, however, involves electrons surrounding the nucleus, and the rate of decay of radioactive isotopes decaying by electron capture might conceivably be affected by changes in chemical structure or pressure or, possibly, temperature. This possibility was first suggested by Segrè,[2] and many experiments testing that suggestion were summarized in Emery[3].

Since the electrons to be captured are primarily from the innermost orbits (lowermost energy levels) of the electrons surrounding the nucleus, most of the experiments summarized by Emery were performed on radioactive beryllium-7, which decays to lithium-7 by electron capture. Neutral beryllium atoms contain just four electrons, so the innermost electrons are not shielded by very many outer electrons, and therefore these inner electrons are more likely to be affected by external factors than would be the case for larger atoms with more electrons. Emery reports changes in the decay rate of Be-7 due to changes in chemical structure amounting to up to 1.8 parts per thousand (0.18 %).[4]

Large changes in external pressure might also affect the inner electrons of atoms, and therefore might affect the radioactive decay rate of isotopes that decay by electron capture. A report of a change in the decay rate of Be-7 of up to 6 parts per thousand under pressure up to 270 kilobars was reported soon after the publication of the review article by Emery.[5] (One kilobar equals 1000 bars; 1 bar is slightly greater than the average atmospheric pressure at sea level on Earth.) The pressure in the interior of the Earth increases with depth, and reaches 270 kilobars at a depth of about 700 kilometers. Rocks undergo metamorphism at depths greater than about 10 kilometers, where the pressure is about 3 kilobars, and the deepest parts of the Earth's crust are at a depth of about 60 kilometers, where the pressure is about 19 kilobars. Therefore, most rocks, and all crystalline igneous rocks, have never experienced a pressure greater than a few kilobars. Since a pressure of 270 kilobars produces a change in the half-life of Be-7 of less than 1%, we may be confident that the low pressures experienced by Earth's crustal rocks would not be expected to produce any observable changes in the decay rate of Be-7.

Recent experiments on the changes in decay rate of Be-7 due to changes in chemical environment or in pressure, using modern equipment with much higher precision, report similar results.[6,7]

So the answer to question #1 above is yes, the rate of decay of radioactive isotopes that decay by electron capture can be changed by changes in the surroundings of the atoms of such isotopes. The changes that have been observed are small, of the order of 1 % or less.

Potassium-40 is the only isotope commonly used in measuring the ages of rocks that decays by electron capture. No report of any experiments attempting to measure changes in the decay rate of K-40 was found in the scientific literature by this author. Liu comments about the probability of any such changes as follows:

"The conversion of K-40 to Ar-40 by electron capture has been widely adopted to date geological events. If the effect of pressure on the decay rate of Be-7 observed in the present study also occurs in K-40, and potassium-containing minerals were subjected to high pressures during their geological history, the ages of these materials determined by the conventional dating method might be overestimated. However since the decay rate of Be-7 increases by about 1 % at 400 kilobars, it would be expected that a similar effect on larger nuclides such as K-40 would be smaller. We would like to note here that, following our experiment on

Be-7, another experiment was performed on rubidium-83, a much bigger nuclide undergoing electron-capture decay. For a nuclide of this size no measurable changes were observed up to 420 kilobar at room temperature."[8]

The errors that might result from changes in chemical environment or pressure in measuring ages of rocks by the potassium-40/argon-40 method, therefore, are unlikely to amount to as much as one percent. Therefore we can confidently conclude that such changes would not lead us into any appreciably mistaken conclusions about the ages of rocks.

The experiments described above, and the consideration of the constancy of fundamental constants discussed in Chapter 19, indicate that there have been no significant departures from constancy in decay rates of radioactive isotopes during the history of the universe.

References

[1]G.T. Emery, "Perturbation of nuclear decay rates," in *Annual Review of Nuclear Science* 22 (1972), 165-6.

[2]E. Segrè, "Possibility of altering the decay rate of a radioactive substance," *Physical Review* 71 (1947), 274-5.

[3]Emery, "Perturbation," 175-96.

[4]Emery, "Perturbation," 177.

[5]W. K. Hensley, W. A. Bassett, and J. R. Huizenga, "Pressure dependence of the radioactive decay constant of beryllium-7," *Science* 181 (21 September 1973), 1164-5.

[6]Chih-An Huh, "Dependence of the decay rate of [7]Be on chemical forms," *Earth and Planetary Science Letters* 171 (1999), 325-8.

[7]Lin-gun Liu and Chih-An Huh, "Effect of Pressure on the decay rate of [7]Be," *Earth and Planetary Science Letters* 180 (2000), 163-7.

[8]Liu and Huh, "Effect of Pressure," 166.

Appendix E
Production of Carbon-14 (Chapter 21)

After neutrons were identified as particles of matter in 1932, investigations of Earth's atmosphere found that there were neutrons there. Since individual neutrons were known to decay into protons and electrons with a half-life of 12 minutes, there must be some mechanism by which neutrons are being produced in the atmosphere, and eventually the mechanism was identified as the interaction of cosmic-ray particles with the components of Earth's atmosphere, primarily nitrogen and oxygen. Further investigation found that carbon-14 is being produced by the interaction of nitrogen-14 with those neutrons in Earth's atmosphere. Soon thereafter, the radiocarbon dating method was developed and put to use.

The investigations of neutrons in Earth's atmosphere included measurements of the number of neutrons found in a sample of known volume or weight of air. Measurements were done at various altitudes, using instruments carried by aircraft and by balloons. Since it was known that the electrically charged cosmic ray particles are affected by Earth's magnetic field, measurements were also carried out on airplane flights at constant altitude over a range of geomagnetic latitudes, that is, latitudes with respect to the magnetic poles of the Earth.

During the same span of history, many measurements were being carried out to determine the probability of the interaction of neutrons with various isotopes of the elements, using neutrons produced by particle accelerators such as cyclotrons, or by nuclear reactors. The probability of the reaction of neutrons with N-14 to produce C-14 has been measured and published.

With information on the number of neutrons in a given quantity of Earth's atmosphere, and information on the probability of the reaction of those neutrons with N-14 to produce C-14, it is possible to calculate the rate of production of C-14 in that quantity of the atmosphere. To get an estimate of the rate of production for the whole Earth, one must also take into account the variation in the number of neutrons with altitude in the atmosphere, and the variation with latitude. For long-term production rates, one must also take into account any variation with sunspot cycle, occurrences of solar flares, and any other potential influences on cosmic ray incidence on Earth's atmosphere. There are significant uncertainties in our knowledge of the extent to which those factors affect the production rate of C-14.

The decay rate of C-14 can be determined by standard procedures for any particular sample, and the average for plants and animals can be obtained by doing such measurements for a sufficient number of samples. However, estimates of the rate of decay of C-14 for the whole Earth also require knowledge of the amount of C-14 in Earth's atmosphere, in Earth's oceans, in the total of all plants and animals on Earth, etc., and also requires knowledge of the total amount of carbon (mostly C-12) in Earth and atmosphere that is in the "carbon cycle," that is, all the carbon in the entire Earth that is actively interacting with the C-14 of the atmosphere. There are significant uncertainties in our knowledge of those quantities.

In describing the radiocarbon dating procedure in his book *Radiocarbon Dating*, Libby published a comparison of his calculated estimates of the production rate of C-14 for the entire Earth with estimates of the decay rate of C-14 for the entire Earth. The values he gives are a production rate of 18.8 atoms of C-14 per gram of total carbon per minute, and a decay rate of 16.1 ± 0.5 decays per gram of total carbon.

In comparing the production rate and decay rate given, Libby stated that "The agreement seems to be sufficiently within the experimental errors involved" to justify the conclusion that the production rate and decay rate are in equilibrium, that is, in a steady state. Libby did not provide a value for the uncertainty in the production rate of C-14 that he had estimated, but he did provide references to the publications on which he had based his calculation of 18.8 atoms of C-14 per gram of carbon. To get some idea of the uncertainty in that value, consider the following:

1. The probability of the reaction of neutrons with N-14 to form C-14 was not given in Libby's discussion, nor was it investigated by this author, but the uncertainty in that probability is certainly larger than zero, and is probably at least a few percent of the value given.

2. The value used by Libby for the number of neutrons in Earth's atmosphere at various altitudes was derived from high-altitude balloon flights, the highest to about 67,000 feet, launched from New York (state) at a geomagnetic latitude of 55 degrees north.[1] The number of neutrons reported there is obtained as the difference between two numbers, each of which has a reported uncertainty of about ± 10% of the value given.

3. The variation with latitude used by Libby comes from one flight of a B-29 from Long Island, NY to Panama, geomagnetic latitude 52 degrees north to 20 degrees north, at a constant altitude of 25,000 feet.[2] The number of neutrons found is the difference between two measurements, each with uncertainty of about ± 10% of the value given.

4. The uncertainties reported in 2) and 3) above are based on counting statistics only. No accounting of additional sources of possible error is published in the resource publications consulted by Libby.

It is clear that there is a considerable uncertainty in the accuracy of Libby's estimate of the production rate of C-14 for the entire Earth, adequately justifying Libby's conclusion that the values of the production rate and the decay rate of C-14 can be considered to be equal within the uncertainties of the quantities used in his calculations.

References

[1] L.C.L. Yuan, *Physical Review*, 74 (1948), 504.

[2] Yuan, *Physical Review*, 76 (1949), 1267-8.

Appendix F
Ages of Rocks in the Grand Canyon (Chapter 25)

We noted in Chapter 25 that the conditions required for rock samples to provide a valid age measurement by the rubidium-87/strontium-87 method are not met by samples of the volcanic rocks from recent eruptions in the area of the Grand Canyon of Arizona. That conclusion has been published in the professional scientific literature as well as in *Grand Canyon: Monument to Catastrophe*.[1] That conclusion deserves some further comment.

Rubidium-Strontium age measurement

Rubidium and strontium are present in various rock-forming minerals as minor or trace constituents. Because rubidium is chemically similar to potassium, rubidium is commonly found as an impurity in minerals containing potassium. Because strontium is chemically similar to calcium, strontium is commonly found as an impurity in minerals containing calcium. Rubidium-87 is the radioactive parent of this pair, with a half-life of 49 billion years, and strontium-87 is the daughter isotope.

Both rubidium and strontium are present in small amounts in terrestrial rocks. The half-life of Rb-87 is very long, so only a small change in the amount of Sr-87 results from the decay of Rb-87, even over very long periods of time. Consequently, the straightforward parent-to-daughter procedures that are useful for potassium-40/argon-40 or uranium-238/lead-206 are useful only for Rb-87/Sr-87 measurements in minerals that contain a considerable amount of rubidium. For other rocks, a graphical solution is employed, using ratios of isotopes, since the ratios of isotopes can be determined very accurately with a mass spectrometer. The graph is constructed in such a way that it depicts an "isochron," meaning "same age."

An isotope of the same element as the daughter of the radioactive parent is chosen as the denominator of the ratios, using an isotope that does not change over time. The ratio of the daughter isotope to that stable isotope is plotted on the vertical axis of the graph, and the ratio of the radioactive parent isotope to the stable isotope on the horizontal axis. For each atom of the parent that decays, an atom of the daughter isotope is formed. As time goes by, the amount of parent isotope in the sample decreases, and the amount of daughter isotope increases, atom for atom. After a time has passed, the plot of ratios will reflect that change. Ideally, a series of samples from the same rock mass are selected that contain different initial amounts of parent isotope; all of them, however, will have the same ratio of daughter isotope to stable isotope when the rock was formed, assuming that the samples come from a source with uniform isotopic composition (such as a reservoir of molten magma that is well mixed). A plot of those initial ratios will lie along a

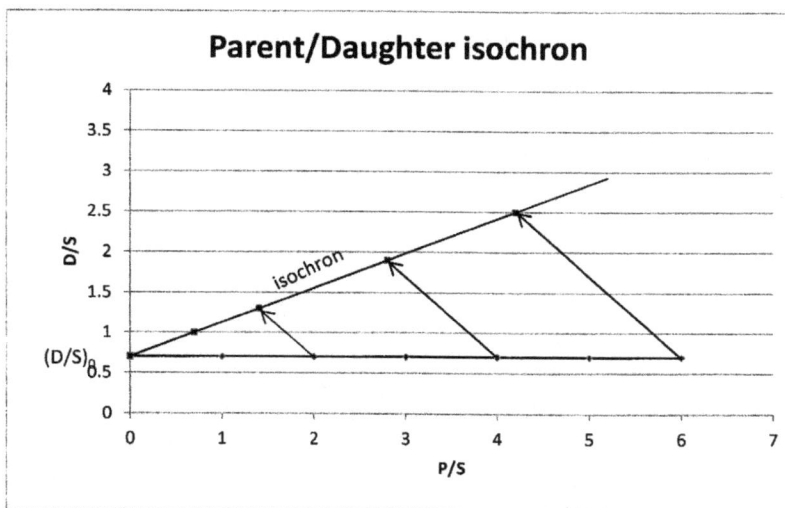

Figure 9. Parent/daughter isochron.

horizontal line. After some time has passed, a plot of the changed ratios will still lie along a straight line, but now on a sloped line, since the amount of change is proportional to the amount of the parent isotope in the sample. This sloped line is called the "isochron." The intercept of the line with the zero vertical axis will fall at the initial ratio of daughter to stable isotope, since it represents a sample with no radioactive parent isotope. The slope of the isochron, together with the known half-life of the parent isotope, is a measure of the time that has passed since the rock was formed, that is, the age of the rock.

The graph shown in Figure 9 depicts the "isochron," on which the points fall after a significant fraction of the parent isotope in each sample has decayed into its daughter isotope. The symbol "D/S" represents the ratio of the amount of daughter isotope to the amount of stable isotope, the symbol "P/S" represents the ratio of the amount of radioactive parent isotope to the amount of stable isotope, and the symbol "$(D/S)_0$" represents the initial ratio of the amount of daughter isotope to the amount of stable isotope, that is, the ratio when the rock was formed. In practice, a number of samples are collected from different places in the rock body of interest, such as a lava flow, or a granite intrusion. The sample may be an individual mineral grain from the rock, or may be a "whole rock" sample, consisting of a mixture of minerals. Any piece of rock with evidence of weathering, or contact with groundwater, or other alteration of the rock would not be acceptable as a suitable sample.

Next, accurate analyses are required. For the rubidium/strontium pair of isotopes, the procedure called "isotope dilution" is commonly used to measure the amounts of rubidium and strontium in the rock sample of interest; this procedure (which we will not describe in detail here) is capable of determining small amounts of rubidium and strontium with an uncertainty of at most 1% of the amount present. The radioactive parent isotope is rubidium-87, decaying to strontium-87, and strontium-86 has been selected as the stable isotope. The ratio of Sr-87 to Sr-86 is measured to very high accuracy with a mass spectrometer. The average fraction of rubidium in the Earth's crust that consists of the isotope Rb-87 is 27.83%; the amount of Rb-87 in the sample being analyzed is calculated by multiplying the total amount of rubidium from the isotope dilution analysis by that percentage. The amount of Sr-86 in the rock sample is calculated in similar fashion; 9.84% of the strontium in Earth's rocks is Sr-86. The ratio of Rb-87 to Sr-86 is determined from those amounts.

Then a graph is constructed with the ratio Rb-87/Sr-86 plotted on the horizontal x-axis of the graph, and the ratio Sr-87/Sr-86 on the vertical y-axis. The age of the rock can then be determined from the graph.

Hopefully, the samples chosen will contain a range of rubidium contents, the wider the range the better, and hopefully the amount of rubidium in the samples will be scattered at intervals along that range (on the graph) rather than being bunched up at or near one end or the other.

The Rb-87/Sr-87 isotope pair began to be used to date rubidium-rich minerals in the 1950's. Improvement in the precision of mass spectrometers and the development of the isochron method led to the use of Rb-87/Sr-87 in rock samples containing lesser amounts of rubidium in the 1960's. Numerous studies have been done in the laboratory and in the field to investigate strontium isotopes in volcanic rocks. Basaltic lava, with higher amounts of iron and magnesium, is thought to be formed from magma derived from melting of upper mantle material, while continental igneous rocks with more silica, potassium, and sodium, and less iron and magnesium, are derived from melting in the lower crust of the Earth.

It has been observed that there is some variability in the initial Sr-87/Sr-86 ratio in recent basaltic lava flows on several islands in the Pacific, although the magma apparently originated from upper mantle material in all cases, suggesting that the source rock is not thoroughly mixed to give uniform isotopic ratios.[2] The authors of that study suggested that the isotopic ratios might be inherited from the source of the magma in Earth's mantle, and that the isochron represented the time since different pockets of magma derived from different regions of the mantle were separated from the convecting mantle itself. But the isochron certainly does not represent the time since the magma was erupted onto Earth's surface to form the lava from which the samples were taken. Such isochrons came to be called "eruptive isochrons."

Also, the initial isotopic ratio of Sr-87/Sr-86 in continental lavas from the western United States has been found to be significantly higher than the initial isotopic ratio in oceanic basaltic flows.[3] The authors attributed the results to differentiation within the mantle material below the continental crust, and they thought that the isochron line along which the points lie in the graph might represent the time since that differentiation occurred.

Furthermore, it is known that some basaltic lava flows contain partially dissolved inclusions of the continental rock material that the lava passes through on its way to Earth's surface, suggesting significant mixing of dissolved continental rock with the basaltic magma, thus distorting strontium isotope ratios. Various scenarios of mixing of source materials, plus alteration by dissolution of the rock through which the rising molten magma passes on its way to the surface, plus changes in isotopic composition resulting from partial melting and partial crystallization within the magma chamber, produce inconsistencies in the results of the Rb-87/Sr-87 age measurements.[4]

North of the Grand Canyon, on the Uinkaret Plateau, there are a large number of volcanic cones, formed by extrusion of basaltic lava. These lie on the upper surface of the Kaibab Limestone, and they were obviously formed in recent times. More than 150 lava flows from these volcanoes entered the Grand Canyon, with cascades that now cover the layered sedimentary rocks that form the walls of the canyon in the vicinity of Toroweap Valley and Whitmore Wash (about Mile 177 to Mile 190 from the Lee's Ferry Bridge). These flows are obviously very recent since many of them lie on the sides of a canyon already eroded deeply into the sedimentary rocks. Some formed dams across the Colorado River, and have since been eroded to the present level of the river.

Various studies have been done on the strontium isotope ratios in other recently emplaced volcanic rocks in western North America, and a rather wide variation in those ratios has been reported in the scientific literature.[5] Consequently, Rb-87/Sr-87 ratios are not considered to be a reliable estimate of the time at which the molten magma was extruded and cooled to form lava rock at or near the surface of the Earth; at most, the isochron might indicate the time at which the magma source of the lava flows was isolated from the isotopically uniform parent material in the upper mantle.

These inconsistencies have led to the conclusion that "The Rb-Sr method has largely been superseded as a means for dating igneous rocks."[6]

Claims from sources promoting the view that Earth is a recent creation

More recently, a new study of the Rb-87/Sr-87 content of these recent volcanic rocks from the lower Grand Canyon and Uinkaret Plateau has been published. The five samples reported lie along a very nice straight line in the Sr-87/Sr-86 versus Rb-87/Sr-86 diagram, with an indicated "isochron" at 1.34 billion years.[7] Like other samples reported from the recent basaltic lava flows in that area, the rubidium content of the samples is very low, so that all the points on the diagram lie quite close to the y-axis (the y-axis represents a rubidium content of zero).

The results of this recent study are then compared with earlier results of measurements of the "age" of the lava flows known as the "Cardenas basalts," exposed in the eastern part of Grand Canyon National Park. The Rb-87/Sr-87 isochron method provides an age of 1.09 billion years for the Cardenas basalts.[8]

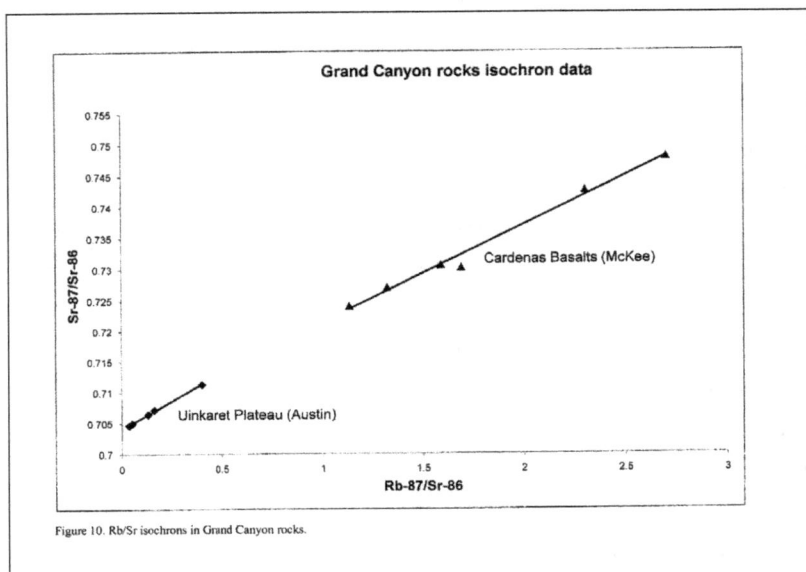

Figure 10. Rb/Sr isochrons in Grand Canyon rocks.

The rubidium content of the six samples from the Cardenas basalts is considerably higher than that in the basalts of the western Grand Canyon, and the spread of points on the Sr-87/Sr-86 versus Rb-87/Sr-86 diagram from the Cardenas basalts covers a much wider range than is covered by the measurements of the samples reported in *Grand Canyon: Monument to Catastrophe*. The isochron line for the Cardenas Basalts and the isochron line from the samples reported in *Monument to Catastrophe* are shown to the same scale in Figure 10.

The Cardenas Basalts are found within the sediments of the Grand Canyon Super Group, far below the surface of the Kaibab Plateau on which the Uinkaret Plateau lava flows were extruded. In terms of geologic position, then, the Cardenas Basalts would be expected to be older than the Uinkaret Plateau flows. Since these results show differently, and since the "age" of the obviously recent Uinkaret flows by the Rb-87/Sr-87 method shows an erroneous age, the result of the same method on the Cardenas Basalts is also unreliable.

The consensus of most of the scientists who have examined Sr-87/Sr-86 values of many volcanic rocks in western North America is that the linear arrangement of points on a Sr-87/Sr-86 versus Rb-

87/Sr86 diagram does not provide a meaningful measure of the time at which the lava flows were extruded, cooled and solidified at or near the surface of the Earth. Rather, that linearity produces a "false" isochron; any linearity represents an isotopic composition inherited from the parent magma, which in turn was derived by melting of material in the upper mantle of the Earth. Although McKee and Noble reported their results in 1976 as the "age" of the Cardenas basalts, I think we can safely conclude that those data, also, do not represent the time of extrusion and solidification of the lavas at or near Earth's surface, but those isotopic compositions, also, were inherited from the parent magma.

Note that the reason that the Rb-87/Sr-87 data fail to provide valid ages of the rocks we have been discussing is that the samples do not meet the stated condition for validity, namely, that the isotopic composition of the body of rock be initially uniform, that is, the same value, throughout that molten magma. The difficulty, however, is not due to any flaw in the Rb-87/Sr-87 method; the difficulty arises because samples of volcanic lava flows do not meet the required conditions for validity.

We have already evaluated the validity of the claim in *Monument to Catastrophe* that the erroneous Rb-Sr results put <u>all</u> measurements of the age of Grand Canyon rocks into question. We will not repeat that discussion here.

References

[1]Steven A. Austin, Ed., "Are Grand Canyon Rocks One Billion Years Old?" in *Grand Canyon: Monument to Catastrophe*, (Santee, CA: Institute for Creation Research, 1994), 111-31.

[2]S.S. Sun and G.N. Hanson, "Evolution of the mantle: geochemical evidence from alkali basalt," *Geology* 3 (1975), 297-302.

[3]C. Brooks, D.E. James, and S.R. Hart, "Ancient lithosphere: its role in young continental volcanism," *Science* 193 (1976), 1086-94.

[4]Beckinsale, R.D., R.J. Pankhurst, R.R. Skelhorn, and J.N. Walsh,

"Geochemistry and petrogenesis of the early Tertiary lava pile of the Isle of Mull, Scotland." *Contributions to Mineralogy and Petrology* 66 (1978), 415-27.

[5]Z.E. Peterman, B.R. Doe, and H.J. Prostka, "Lead and strontium isotopes in rocks of the Absaroka volcanic field, Wyoming." *Contributions to Mineralogy and Petrology* 27 (1970), 121-30.

[6]Alan P. Dickin, *Radiogenic Isotope Geology*, (Cambridge, UK: Cambridge University Press, 2005), 43.

[7]Austin, "Grand Canyon," 124.

[8]Edwin H. McKee and Donald C. Noble, "Age of the Cardenas Lavas, Grand Canyon, Arizona," *Geological Society of America Bulletin*, 87 (1976), 1188-90.

ABOUT THE AUTHOR

Clarence Menninga was born in rural Iowa, earned the B.A. degree from Calvin College, the M.A.T. from Western Michigan University, and the Ph.D. (Chemistry) at Purdue University, plus extensive study in geology at Michigan State University. His employment history includes analytical chemistry in industry, teaching chemistry in high school, and research in U.S. Government research laboratory. He was appointed to the Faculty at Calvin College in 1967 to initiate the program in geology at the College, and he retired from full time teaching in 1990. He has been a participant in archeological work in northern Jordan. He has published a few professional papers in nuclear and radiochemistry and in archeology. He has spoken and written extensively on topics in the relationship of science and Christian faith, including coauthorship of the book *Science Held Hostage*. He resides in Grand Rapids, Michigan, and continues to participate in activities of the GEO Department at Calvin College.

www.ingramcontent.com/pod-product-compliance
Lightning Source LLC
Chambersburg PA
CBHW060229050426
42448CB00009B/1357